For information, address:
Crossfuze, 500 Seneca Street #620, Buffalo, New York 14204

ISBN: 978-0-9993184-2-3

Visit www.crossfuze.com

Crossfuze
Enterprise Services That Win

Table of Contents

Introduction 6

Pillar 1: Delivering HR to Employees Through Services 10

Pillar 2: Selling Your Sales Team On Next Generation Customer Service Management 18

Pillar 3: Managing Project Portfolios with ESM 26

Pillar 4: Adapting the ESM Model to Service Maintenance and Operations 34

Pillar 5: Managing Assets and SLAs Through ESM 42

Pillar 6: Planning and Executing Marketing Strategy Using ESM 50

Pillar 7: Enhancing Financial and Budget Management with ESM 58

Pillar 8: Improving Management of Software Development Using ESM 66

Pillar 9: Infusing Cybersecurity Operations with the ESM Model 74

Pillar 10: Integrating and Coordinating ESM Across Business Functions 82

Final Thoughts 90

Afterword 92

About Crossfuze 97

QUICK TIP KEY

- Innovation
- Vision Fulfillment
- Team Impact
- Time Management
- ServiceNow Insight

Foreword

If you're like most enterprise leaders, you're struggling to realize the promise and the potential of digital transformation. Perhaps you can't find a way to get your digital transformation ideas on the C-suite agenda. Maybe you're at a loss for words to explain why a colleague in HR or sales should adopt ServiceNow. Or you might just be wondering why your own ServiceNow transformation plans aren't gaining any traction. This book is here to set you on a path to success.

The *10 Pillars of ServiceNow Digital Transformation for Non-IT Leaders* is written to help every forward-thinking leader in your organization bring clarity and focus to his or her ambitious vision for ServiceNow. The book will jump-start strategic planning initiatives that will expand the reach and impact of ServiceNow to departments across your organization. In fact, you can parachute into any of 10 chapters focused on a particular department or division—or read this book from cover to cover. No matter how you consume this book, remember to pay its knowledge forward by passing the book to colleagues in other departments. Allow this book to help you break down the traditional silos in your organization that are blocking your path to digital transformation. You will be so glad you did.

Introduction

Among the ranks of the C-suite, it is widely assumed that executing on digital transformation is a core competency. Business leaders, after all, should be able to seamlessly roll out next-generation technologies and approaches across the enterprise, and understand the theory and strategy behind digital innovation.

Unfortunately, this assumption is incorrect. Studies consistently show that only a small fraction of enterprise leaders possess a truly transformational mindset. A 2015 study by the consulting firm Harthill found that just 8% of executives share the characteristics of transformational leaders. The reality is that most members of the C-suite aren't receptive to the type of risk-taking and bold thinking that are required to deliver digital transformation; they don't want to jeopardize the career they've worked so hard to create, and prefer the status quo to major disruption.

Compounding the challenge is that even when enterprise leaders attempt to execute on digital transformation, they preside over initiatives that tend to languish, falter, and fail. A recent study by Wipro Digital found that half of all companies are failing to execute on their digital transformation strategies, with only 4% realizing the first half of their digital investment in the first year. These failures collectively add up to a crisis of epic proportions. Leaders are blowing through their budgets and yet struggling to get results. And as the marketplace changes around them at warp speed, these leaders are failing to build momentum to remain current and relevant.

Enterprise leaders tend to be reluctant to embrace digital transformation—and to falter when they do.

One of the defining hallmarks of digital transformation is the opportunity to reimagine the delivery and management of business services across all departments and divisions. Known as Enterprise Service Management (ESM), this transformative approach involves streamlining and automating workflows across the enterprise. ESM enables businesses to drive new, bottom-line efficiencies, advance long-term business strategies, and preserve the business's competitive edge. And best of all, every business stands to benefit from ESM transformation. A 2016 study by McKinsey found that companies can automate at least 30% of their activities across a majority of departments by taking advantage of digital transformation technologies already available to them.

Businesses across all industries can benefit from the Enterprise Service Management approach to digital transformation.

In recent years, ServiceNow has been identified by Gartner as the industry-leading platform for IT service management, and has become the go-to technology for delivering and managing services across the enterprise, from HR to sales to operations and finance. While ServiceNow offers enormous transformative potential, simply buying a ServiceNow license is no guarantee of ESM success. For ServiceNow to be effective, departments need to understand how ESM is applied in a dynamic, fast-paced workplace setting. That means senior leadership has an obligation to communicate with various departments at their level—that is, to initiate in-depth conversations about how ESM supports the specific needs, priorities, and goals of individual departments. It is through tailored messaging that each department of the organization buys into ESM's value proposition and rallies around ServiceNow digital transformation.

At Crossfuze, we understand from firsthand experience that the most effective, lowest-risk way for businesses to achieve ESM transformation is for department leaders to buy into a common vision focused around enhanced efficiencies, streamlined automation, and increased profitability. Unfortunately, department leaders in most organizations tend to take a siloed approach to digital transformation initiatives, and fail to understand how the parts add up to the whole. Siloed leaders focus primarily on their own internal priorities and metrics—not on goals, priorities, and outcomes for the enterprise as a whole. Consequently, digital transformation tends to move unevenly and inefficiently through organizations. Digital transformation initiatives slow down meaningful change instead of driving it. In fact, the more fragmented and disconnected the organization is, the more that leaders fall victim to a lack of coordination and missed synergy opportunities among departments.

Enterprises maximize success with a ServiceNow digital transformation when every department moves in lockstep toward a common vision.

That's why the hard work of rallying an organization around a digital transformation initiative as important as ServiceNow cannot be left only to IT leaders. All leaders must work together to ensure that ServiceNow digital transformation permeates every inch of the organization, and firmly embeds itself in the cultural fabric of the organization. The goal of the *10 Pillars of ServiceNow Digital Transformation for Non-IT Leaders* is to get these conversations among department leaders started, to close gaps in C-suite understanding of how to optimally position ServiceNow for enterprise-wide implementation success.

In this book, we'll introduce you to the nuts and bolts of ServiceNow digital transformation across many disparate departments and service units—what it means for each unit, what it looks like when implemented, and the instrumental role that non-IT leaders must play to achieve transformative success. We'll arm you with tips and advice to help you speak the same language of digital transformation, and teach you how to stay focused on the right issues and questions. You'll gain a clear roadmap for how to maximize the impact of ServiceNow, and how to solidify your digital transformation successes with speed, agility, and confidence.

This book will teach every leader in your organization to speak the language of ServiceNow digital transformation.

Each of the 10 pillars of ServiceNow digital transformation focuses on one traditionally siloed department within the organization, and offers actionable insights and guidance to help each department relate to and embrace a ServiceNow digital transformation. We are confident that over time, ServiceNow can work its transformative magic for every major service unit in your organization. We look forward to teaching the language of enterprise services and ESM digital transformation to every department in your organization, so all of you can craft common goals and plans for ServiceNow that will optimize your workflows and maximize profitability.

PILLAR 1

Delivering HR to Employees Through Services

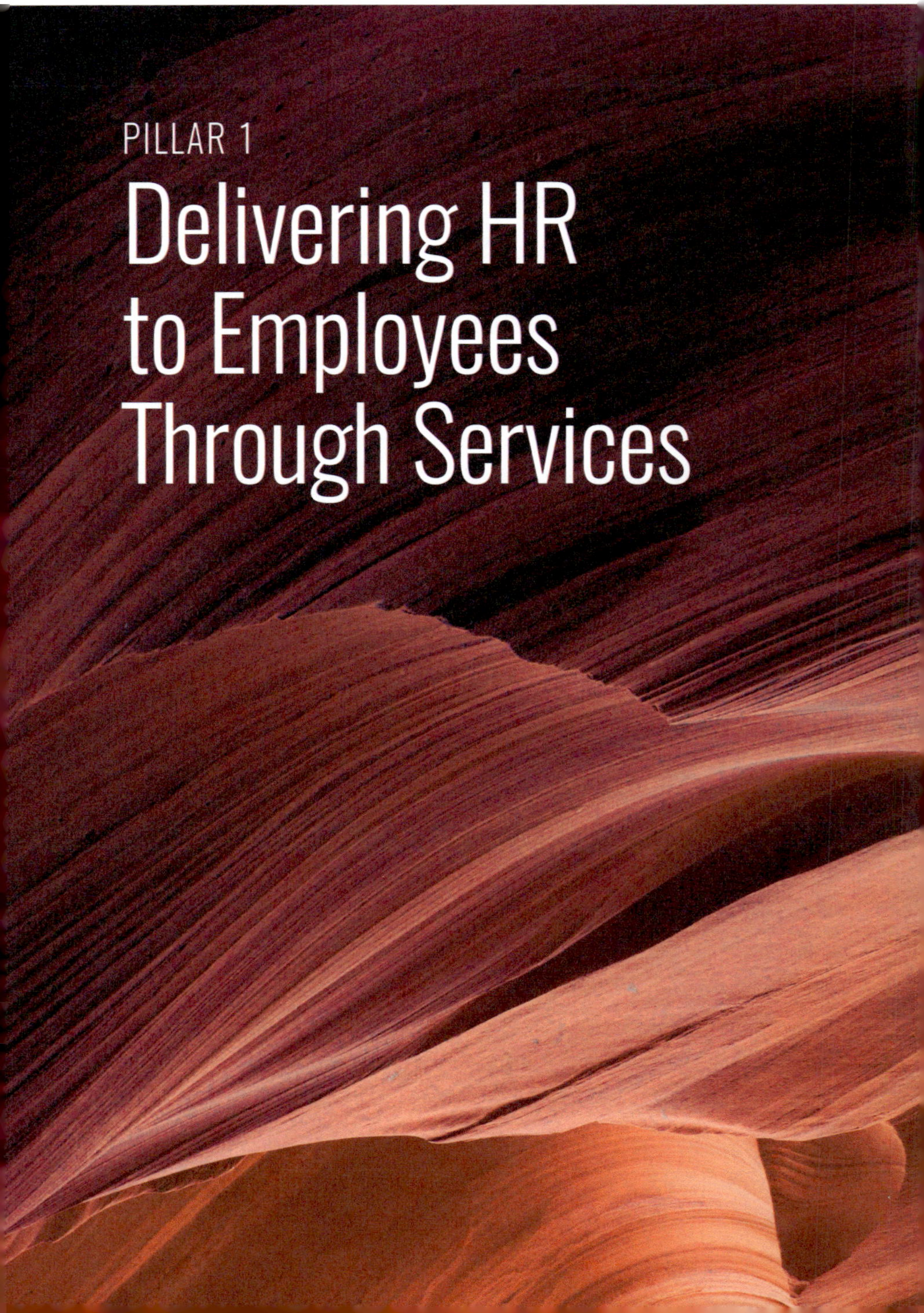

MEET SIMON

Simon is the senior vice president of human resources for an organization that has been growing in leaps and bounds. His department was set up at a time when the company was much smaller, and it's now unable to keep up with demand. Every day, Simon must put out fires stemming from miscommunication and inconsistent communication on his own team. He deals with everything from backlogged onboarding processes to employee-initiated requests that have fallen through the cracks and strategic initiatives that seem to be permanently on hold. Simon recognizes the value of bringing better structure and standardization to HR workflows and processes, but at the same time, he doesn't see a viable path forward to realistically implement changes of this magnitude. Indeed, he doesn't even know exactly what to ask for and how to ask for it. Furthermore, Simon understands that before he can ask for any funding and resources to make sweeping changes, he needs to demonstrate that he has brought accountability and discipline to his team. Consequently, Simon is paralyzed by fear, and declines to rock the boat and disrupt the status quo in any meaningful way.

The challenges that Simon faces are endemic in the world of HR. HR leaders are under constant pressure to meet more service requests from more employees with greater speed and agility, and to cultivate a workplace culture that enhances employee satisfaction, productivity, and retention. A recent study by the Society for Human Resource Management found that every time an employee leaves, the average company spends the equivalent of six to nine months of their salary to find and train a replacement. With such high stakes attached, HR leaders are searching for solutions that can lower costs, make employees want to stay, and advance long-term workplace culture goals. HR leaders are increasingly turning to digital transformation solutions like ServiceNow to achieve their goals. A 2017 Deloitte survey found that one in three HR teams has embraced some form of artificial intelligence technology to deliver HR solutions, and 41% are actively building mobile apps to deliver HR services.

ServiceNow can help HR streamline and automate delivery of services.

Although ServiceNow has its origins in IT, the platform is a cloud-based solution for managing the delivery of all types of services across an enterprise, including HR functions. That makes ServiceNow a prime candidate to automate routine HR processes and streamline day-to-day management of HR resources. Specifically, you'll be able to standardize employee onboarding processes and build self-help portals, enabling you to reduce the time spent on routine processing and answering commonly asked questions. With all of this freed-up time, your HR team will be able to focus on culture-changing workplace initiatives. Let's explore five essential strategies for using ServiceNow to deliver and manage HR services in a way that drives bottom-line efficiencies and supports enterprise-wide digital transformation:

CHA-CHING

Every time an employee leaves, the average company spends the equivalent of six to nine months of their salary to find and train a replacement.

SOCIETY FOR HUMAN RESOURCE MANAGEMENT

1.1 CREATE CONSUMER-GRADE SERVICE EXPERIENCES:

With the technological revolution has come a plethora of sleek, user-friendly web services and devices that have redefined how we connect, engage, and complete everyday tasks. Employees expect to find these same intuitive features and services when they turn to their HR department. ServiceNow has been designed to deliver these services using out-of-the-box, plug-and-play solutions that require minimal customization and coding. You can create self-help knowledge centers, enable employees to instantly connect with an HR representative through web chat, and allow employees to connect to all of these features anytime and anywhere through their personal devices. These consumer-grade service experiences will reduce workloads for HR representatives, increase employee satisfaction, and give HR a reputation as an innovator that cares deeply about the employee experience.

1.2 STANDARDIZE HR SERVICE REQUESTS:

Common HR processes like employee onboarding and benefit changes consume considerable resources when they are completed in a manual, one-off fashion; these processes also are vulnerable to mistakes and errors. ServiceNow provides frameworks and systems that allow you to seamlessly standardize common HR services requests—and to automatically generate standardized, comprehensive records of these activities. In effect, ServiceNow automates the consistency with which common HR services are delivered. Furthermore, such standardization won't just encompass the day-to-day tasks of HR professionals; it also will extend to the employee being served, including the ability to automatically prompt the employee when to take action and to automatically send an appropriate set of instructions for completing these tasks. While HR services are traditionally delivered in a rather haphazard, inefficient way, ServiceNow harnesses the power of automation and standardization to replace clunky manual processes with optimized delivery approaches.

ServiceNow supports creation of standardized, consumer-grade HR service experiences.

1.3 ROUTE CASES TO THE HR REPS BEST EQUIPPED TO HANDLE THEM:

Within every organization are subject-matter experts and people with a natural proclivity for certain work tasks. Your HR department may have an intuitive understanding of who is best qualified to handle certain types of tasks, but any redirection of tasks to these individuals likely is done manually and on a whim. ServiceNow was designed to automate this pain point, because its origins in IT have underscored the importance of being able to automate case routing to improve speed, efficiency, and quality of outcomes. Through intelligent case routing, ServiceNow "learns" the priorities and preferences of the organization as it accumulates more and more usage data. In this way, ServiceNow will keep getting better and more efficient at routing requests over time.

ServiceNow supports intelligent case routing to maximize use of expertise and talent.

1.4 USE SELF-SERVICE PORTALS TO GIVE TIME BACK TO EMPLOYEES:

It's no secret that many requests that HR spends precious time processing are routine and don't require a trained HR professional in the first place. Moreover, many employees would prefer to complete the task themselves; indeed, the Internet era has taught us that the best way to complete a task is to Google it. ServiceNow rises to the occasion by providing the infrastructure to build self-service portals that contain answers to commonly asked questions. The portals even feature a Google-style search bar with suggestive-text capabilities, seamlessly extending Google's intuitive user experience to many common HR service requests. When employees are able to get answers and complete HR-related functions without any help from a live HR agent, it will dramatically reduce caseloads, and enable HR professionals to focus on higher-value projects and more forward-thinking, strategic goals. Simply put, self-service portals change the lives of both employees and HR professionals for the better.

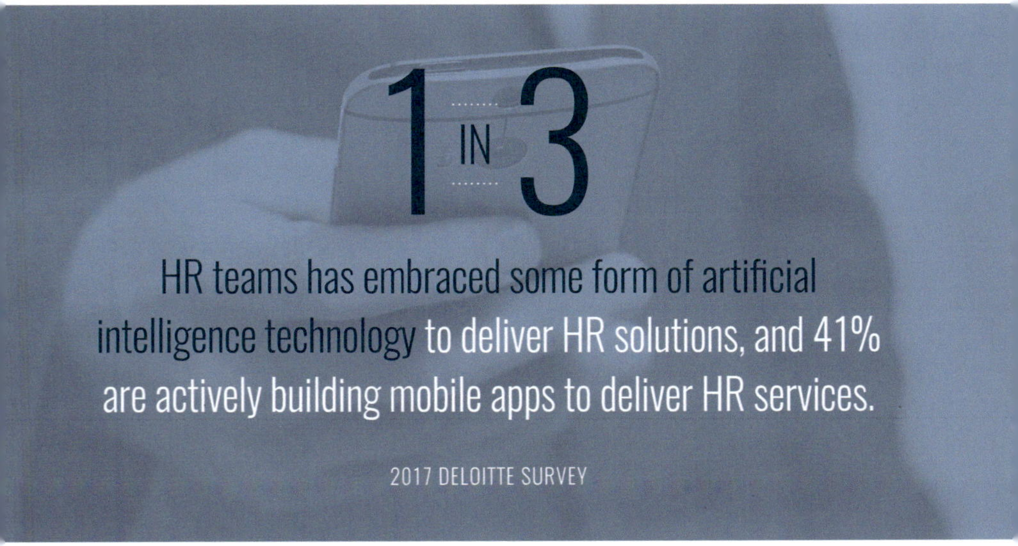

1.5 SERVE UP PERSONALIZED CONTENT TO PARTICULAR TYPES OF EMPLOYEES:

With the buildout of self-contained HR knowledge centers comes the opportunity to target certain groups of people with certain HR content at specific times. If your company undergoes a change in benefits that only affects employees with enrolled dependents, for example, you could push out notices, FAQs, use cases, and a link to the web enrollment portal to only impacted employees. You also could configure reminders to pop up automatically on the computers of just these employees. There is no limit to how you can serve up content based on location, role, job type, and so forth. Ultimately, this targeted, personalized content will increase participation, response, and click-through rates, and minimize the amount of redundant, non-targeted HR content that employees see.

ServiceNow's HR automations free up employee resources and enable delivery of more targeted, personalized content.

HR departments waste too many of their precious resources and talents completing tasks and requests that have the potential to be automated. When an HR department is able to take advantage of ServiceNow to free up time for value-add projects, it's a win-win situation: The department is perceived as employee-friendly and forward-thinking, thus helping the entire organization to succeed with digital transformation initiatives. To maximize your chances of achieving these outcomes with ServiceNow, focus on creating consumer-grade HR services, standardizing common HR tasks and requests, using machine intelligence to route HR cases to personnel best equipped to handle them, reducing caseloads through implementation of self-service portals and functions, and serving up targeted content to particular types of employees.

“

Today, HR’s focus has shifted toward building the organization of the future. Companies are hiring young, digitally savvy workers who are comfortable doing things themselves and sharing information in a transparent way. They want an integrated, digital experience at work—one designed around teams, productivity, and empowerment—and HR is expected to deliver it.

DELOITTE INSIGHTS

PILLAR 2

Selling Your Sales Team On Next-Gen Customer Service Management

MEET ANGELICA

Angelica is the newly minted vice president of customer experience for a software company, and the first big task on her plate is to improve her company's lackluster customer retention rate. The company already uses Salesforce to manage customer relationships, but this solution isn't moving the needle. To get to the bottom of this problem, Angelica begins by interviewing her customer service leaders and reviewing the files of customers that have jumped ship. It doesn't take long for Angelica to sense a major disconnect between the sales team and their counterparts in IT and dev-ops. Although the IT folks are supposed to be working closely with sales to close deals and fulfill customer requests, Angelica documents multiple instances in which these teams do not follow up on requests from customers to modify and customize core product features. Even more troubling, she discovers instances in which requests from customers to modify terms of service for sales contracts have slipped through the cracks. Angelica recognizes that this customer service nightmare must be remedied, but she doesn't know exactly what to ask of IT or how to collaborate with them on an effective solution. In essence, Angelica's own shortcomings as a problem-solver perpetuate the organization's embarrassing customer retention rate.

Angelica's struggles to improve customer satisfaction and deal-closing rates in a competitive sales environment aren't unique. Many customer experience leaders struggle with how to approach IT to help solve the issues at hand, and end up applying Band-Aid solutions to mask the problem. These temporary solutions fail to deliver long-term value and don't support enterprise-wide digital transformation. Meanwhile, customers are acutely aware that these shortcomings ultimately manifest as sub-optimal customer service experiences. A 2017 study examining how business professionals perceive B2B salespeople found that nearly two-thirds rate these salespeople as being average or poor, with the two most common reasons being that salespeople don't tailor their sales pitches and don't listen or understand buyer requirements. Even among companies that use Customer Relationship Management (CRM) solutions like Salesforce, these negative perceptions persist—and they all boil down to a problem with customer service. The challenge is that while CRM is a powerful tool for bringing in new business and maintaining ongoing sales relationships to grow accounts, CRM does not offer preloaded features and workflows for resolving customer service issues, nor does it connect sales to experts in IT, dev-ops, and customer service who are optimally positioned to help resolve these service issues.

You can't close the disconnect between customer service and IT with Band-Aid solutions.

As any sales and marketing professional understands, the battle to achieve high customer satisfaction scores is lost and won on the organization's ability to deliver a superior, full-cycle customer experience. That's why ServiceNow has been designed to pick up where Salesforce and other CRM solutions have left off. ServiceNow brings structure and regimented workflows to customer service experiences that otherwise would be very ad-hoc, unpredictable, and uneven. Moreover, ServiceNow as a customer management platform is designed to complement and enhance—not compete with—Salesforce and other CRM solutions. Let's explore four essential strategies for using ServiceNow to enhance the all-important customer experience that your sales team is working so hard to perfect:

servicenow

ServiceNow turns customer service into customer success in a way that even Salesforce cannot.

2 OUT OF 3

business professionals rate B2B salespeople as being average or poor, with the two most common reasons being that salespeople don't tailor their sales pitches and don't listen or understand buyer requirements

2.1 GET OTHER DEPARTMENTS TO THINK FROM THE CUSTOMER'S POINT OF VIEW:

Your sales team is constantly thinking about how to appeal to customers so they'll open their wallets and spend more with you. This team has perfected the art of identifying pain points, and then solving them with attractive offers, different packaging and price plans, and engaging sales pitches. That said, your sales team is not the only department offering solutions to customers: IT, dev-ops, and customer service also can offer a very different set of solutions to common pain points—but only if these teams are immersed in the customer mindset and focused on creating an exceptional customer experience from start to finish (and beyond). ServiceNow integrates with the CRM solution that your sales team is already using, enabling other departments to see the world from the customer's point of view. The solutions that other departments will come up with will be very different than what your sales team will develop; IT, for example, is most likely to identify opportunities to build new automations and customer-friendly tools.

ServiceNow keeps your IT and dev-ops teams thinking in the customer mindset.

2.2 TAKE ADVANTAGE OF SOPHISTICATED CUSTOMER SERVICE MATRICES:

CRM solutions like Salesforce are designed to drive sales above all else, and they are highly effective at that task. When customers need more than just a steady stream of sales pitches and marketing messages, however, CRM falls flat on its face. CRM is not designed to provide a regimented, consistent level of support and service to customers that require it. ServiceNow's Customer Service Management (CSM) application, by contrast, was expressly created to provide this enhanced level of customer service, offering a workflow-driven approach to resolving customer cases. These customer service matrices seamlessly cut across multiple departments, and include rule-based case routing to ensure that the best departments and individuals for the job are assigned to work on each case.

2.3 TIE CUSTOMER SERVICE CLOSELY TO SALES:

Closing deals and retaining customers involves much more than just a skilled sales team. For the customer to feel good about signing on the dotted line, you need to provide stellar customer service through the full sales lifecycle. Rather than count on each member of the sales staff to work individually with the customer service team to provide this high level of service to customers, ServiceNow can coordinate these processes using standardized automations, ensuring your customer service team is consistently and methodically looped into sales conversations. Once customer service is looped in, customer requests and concerns can be automatically assigned to the representative best equipped to resolve the case. Additionally, all of the actions that the customer service representative takes are automatically recorded in ServiceNow. This means that, for example, if customer service asks the customer for a shipping address, then every other agent who touches the account will have access to that shipping address, so the customer isn't bothered two or three times to provide the same information.

Customer service opportunities won't slip through the cracks with ServiceNow's reliable case management capabilities.

2.4 SYNC SERVICENOW WITH YOUR CRM:

To manage the sales lifecycle with your customers, your sales team will still need a CRM solution like Salesforce. Your goal is to sync ServiceNow with your CRM tool to create a continuous, bidirectional exchange of data. When your sales, IT, dev-ops, and customer service teams are all working off the same customer data—in essence, sharing a holistic view of the customer—they are able to better communicate and collaborate to meet the customer's needs and manage each sale through its full lifecycle. The continuous bidirectional syncing is critical: For example, if the sales team is waiting on dev-ops to modify a component of a software package, the sales and customer service teams needs to be kept abreast of the status of this work order and to know as soon as the work is completed, so they can notify the customer and then follow up later to ensure the software modification has met the customer's specifications.

ServiceNow is designed to enhance Salesforce and other CRMs, not replace them.

Providing high-quality customer service to customers involves more than just the talents of the sales team that manages the sales lifecycle. Increasingly, customer service depends on forming collaborative relationships that link sales, IT, and dev-ops to the customer service team itself. ServiceNow fosters these connections in a systematic, automated fashion, paving the way for digital transformation across the enterprise. With ServiceNow, all departments are able to think in the customer mindset, take advantage of structured customer service matrices, tie customer service closely to sales, and sync ServiceNow with CRM platforms.

PILLAR 3

Managing Project Portfolios with Enterprise Service Management

MEET ATTICUS

Atticus is the newly appointed head of his company's Project Management Office (PMO). Atticus was put into this role because of his reputation for quickly assessing and turning around underperforming business units. Improving the PMO, however, is unlike any challenge Atticus has faced to date. The PMO hasn't achieved the integral, essential role within the company that it should have. Instead, the PMO operates largely on the fringes, and its decision-making authority is tainted by too much influence from the C-suite, which unfortunately is known for shifting priorities and uneven, inadequate funding allocations. Atticus understands that his PMO is the antithesis of what a PMO should be, and he's prepared to propose sweeping changes to the PMO's role and functioning. But first he wants to have a best-practices framework and processes in place for facilitating this daunting transition. How can he use Enterprise Service Management (ESM) principles to maximize his chances of success?

Project managers and PMOs just like Atticus struggle to bring consistency, transparency, and discipline to Project Portfolio Management (PPM). They run into obstacles getting the attention and focus from the C-suite that they need, they lack the proper tools to communicate strategically, and they default to doing too many things manually in a one-off, disconnected fashion. A 2015 University of Houston study examining barriers to PPM implementation found that nearly a third of PPM professionals report being unable to implement a PPM program, and they cite organizational culture as the single biggest challenge to effective management of project portfolios. Furthermore, problems with organizational culture are more than twice as likely to be cited as problems with resource allocations, and seven times more likely to be cited as budget constraints, according to the University of Houston study.

Building an effective Project Management Office is a pervasive problem for organizations.

While entrenched company culture and legacy workflows will always be barriers to achieving PPM success and digital transformation, these challenges do not mean that organizations struggling with PPM should toss in the towel and give up. To the contrary, the ESM model places the dynamic, fast-moving pieces of PPM into structured workflows that are simple to follow and intuitive to visualize. Through ServiceNow's PPM application, project managers get access to powerful tools for managing both resources and demand, enabling PPM to support enterprise-wide transformation. Let's explore five essential strategies for using ServiceNow PPM to bring discipline to your project management workflows, and for elevating the prominence and respect for PPM across your organization:

ServiceNow's Project Portfolio Management app provides structured workflows to easily manage resources and demand.

3.1 ALIGN PROJECT PORTFOLIOS TO STRATEGIC BUSINESS GOALS:

When your C-suite turns its attention to PPM, what these executives really care about is how project portfolios can be managed in ways that strategically advance the organization's long-term goals. The C-suite tends to parachute in for major status updates and strategic planning sessions, but is largely disconnected from the day-to-day challenges of project management. Thus, when you want the C-suite to pay attention, you want to take advantage of ServiceNow PPM to communicate progress, goals, and ongoing challenges. ServiceNow PPM is optimally suited to present data visualizations in ways that are simple and intuitive to understand, and that ultimately are relatable to the C-suite. For example, ServiceNow PPM is pre-configured to present budget and labor constraints alongside project value, and to automatically sort and rank every project according to risk, return on investment, and cost.

ServiceNow PPM presents data visualizations that are simple and intuitive for the C-suite to understand.

3.2 ENGAGE NON-PROJECT MANAGERS WITH STANDARDIZATION AND AUTOMATION:

For PPM to be effective, non-project managers need to understand and engage with PPM processes. Unfortunately, the reality is that the nuances of PPM workflows matter little to anyone who isn't a project manager. To minimize the amount of mental bandwidth required for these folks to help you, you want to standardize and automate as many PPM workflows as you can—and make them as intuitive as possible for your non-project management audiences. ServiceNow, which has its origins in the regimented and standardized world of IT service management, is optimally primed to inject these qualities into your core PPM processes. After all, ServiceNow is about increasing employee engagement and productivity—these are characteristics that translate well to the PPM environment.

3.3 ALLOW PROJECTS TO BE EVALUATED, SCORED, AND RANKED:

The hard work of demand management has traditionally been confined to static, manually generated spreadsheets that are either too complex for anyone outside the PMO office to understand, or not encompassing enough to furnish non-project managers with the key information they need to make informed decisions. Consequently, everyone who touches these spreadsheets struggles to evaluate, score, and rank projects in meaningful, consistent ways. ServiceNow PPM solves this perpetual pain point with expertly designed templates for reviewing and rating projects, as well as powerful visualizations that present project metrics and other relevant data in side-by-side, overview-style formats. Among the features built into ServiceNow PPM are configurable bubble charts, trend analyses, and timeline-based visualizations. These features optimize prioritization processes and ensure all of the right stakeholders understand and become engaged with demand management processes.

servicenow

ServiceNow PPM offers expertly designed templates for reviewing and rating projects—and for fostering stakeholder engagement.

3.4 USE PPM ANALYTICS TO DRIVE BETTER DECISION-MAKING:

Project managers often find themselves engaged in a battle to bring standardization and consistency to departments that want to do things their own way. To rally everyone around common project management principles and goals, you want to be able to show every department that your approach leads to more intelligent decision-making. ServiceNow PPM is your strongest asset in this regard: The application is powered by sophisticated analytical capabilities that will guide you in making objective, transparent decisions and, in the process, will link these decisions to your bottom-line profitability. For example, if project budgets are unexpectedly slashed, ServiceNow's PPM analytics will explain how to optimally slim down your project portfolios and minimize disruptions to operations and productivity. Ultimately, ServiceNow PPM can help you demonstrate to your entire organization that a uniform, transparent, disciplined approach to project management is worth sacrificing departmental autonomy and independence to achieve.

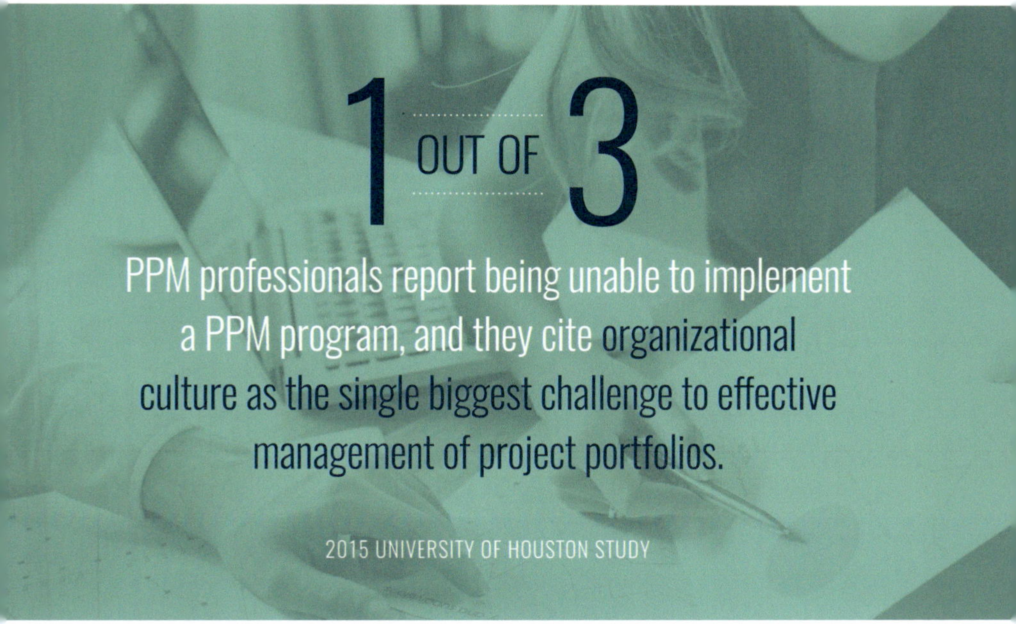

3.5 NEVER STOP PUSHING FOR CONTINUOUS PROCESS IMPROVEMENT:

When a PMO isn't learning and changing, it isn't growing and maturing. And when a PMO isn't growing and maturing, it becomes complacent. Failure points develop without warning, negatively impacting the PMO's effectiveness and creating a domino effect of impacts on other processes and outcomes across the enterprise. The only way to know your PMO is maturing is to systematically and continuously evaluate progress. ServiceNow PPM is designed to empower you to assess progress methodically and quantitatively, with automated reports and visualizations with which to view KPIs and other performance metrics across your organization. By always knowing exactly where you stand on your journey to PMO maturity, you'll be able to optimally allocate your limited resources and energies.

Your PMO can use ServiceNow to methodically and quantitatively assess progress to PMO maturity.

We all endeavor to build a mature PMO that is fully aligned with corporate strategy and embedded deep in the fabric of the organization. Although the journey to build a PMO that is mature enough to support digital transformation takes time and diligence, you can jump-start your progress by using ServiceNow PPM to build respect for and deference to an enterprise-wide PPM philosophy. With ServiceNow PPM at your side, you'll be able to reliably align projects to strategic business goals, create standardization and automation to more effectively engage non-project managers, enable projects to be systematically assessed, make intelligent decisions with PPM analytics, and set a high bar for continuous process improvement.

PILLAR 4

Adapting the ESM Model to Serve Maintenance and Operations

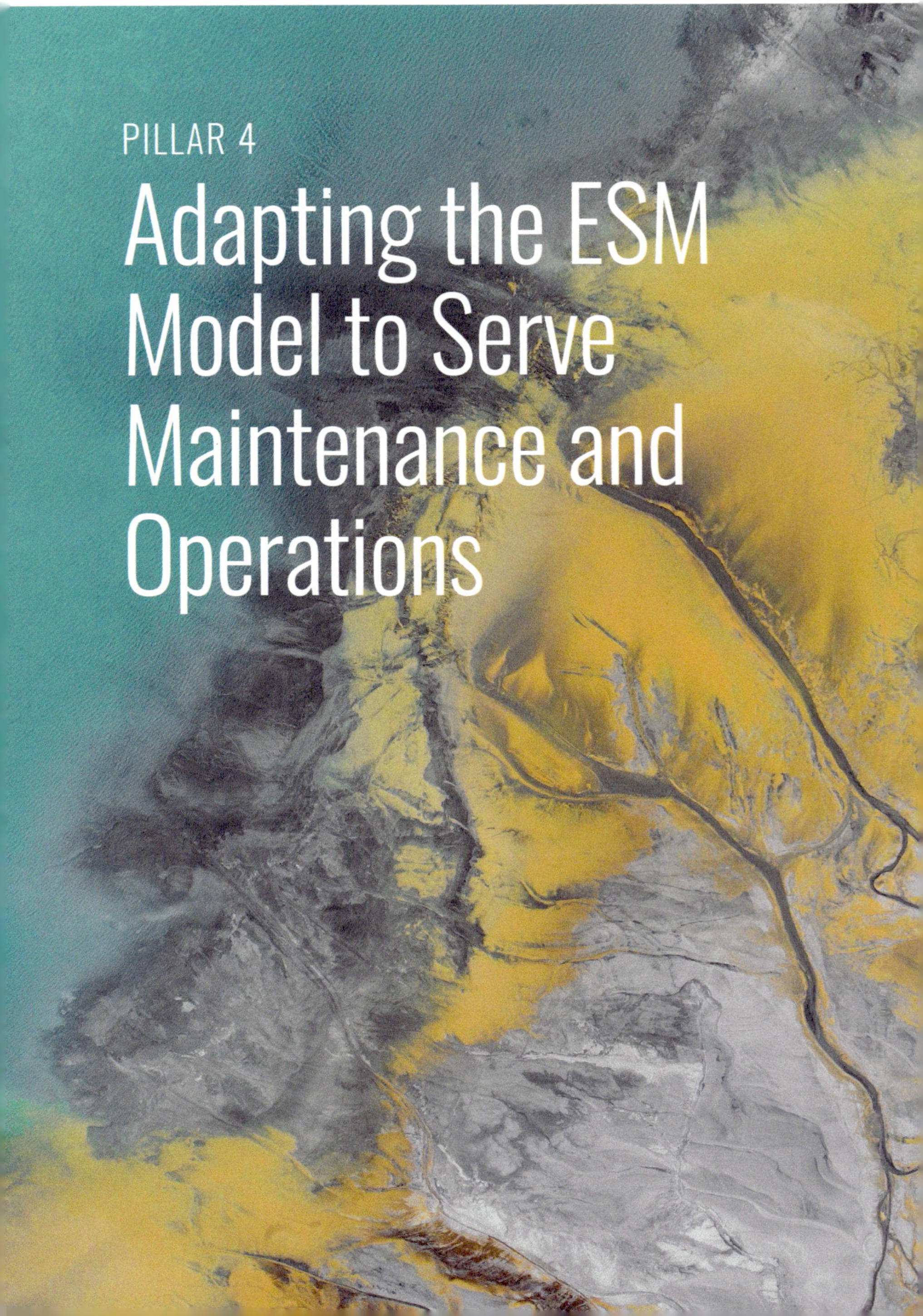

MEET DEVON

Devon is the long-time vice president of operations for a company that is launching a new corporate sustainability initiative. The C-suite has tasked Devon with identifying zero-cost ways for the company to hit more ambitious targets for energy consumption and waste reduction. To meet these targets without any additional hires, Devon recognizes that he will need to free up existing resources across maintenance, operations, and facilities. These service units will need to streamline and reduce manual processes, create more self-service options, and more consistently track and prioritize their workloads. While Devon understands what needs to be done at the conceptual level, he is nervous about execution. Maintenance, operations, and facilities teams in his workplace largely manage requests and track performance in an ad-hoc manner, relying on static spreadsheets and paper logs. He begins exploring an all-encompassing work management solution as a best-practices, universally applicable solution. Devon determines that he wants a system that can structure operational workflows and streamline management of service requests, as well as that can plug seamlessly into his company's existing infrastructure. ServiceNow pops up on his radar, and he's intrigued. Why is ServiceNow—a system that has historically been used by IT departments—the ideal solution to meet his needs?

Many operational leaders just like Devon are acutely aware of the importance of modernizing and streamlining the delivery of their enterprise services to drive profitability and support digital transformation initiatives. A recent study of public utility companies by McKinsey & Company found that optimized plant maintenance, along with better management of spare parts and fuel, can increase earnings by 6.6%. Furthermore, using enterprise resource planning to optimize operations at headquarters can lead to an additional 1.3% increase in earnings. Although these efficiencies seem modest, they can represent crucial make-or-break points for meeting vital cost-savings and corporate sustainability targets.

Maintenance and operations are under pressure to modernize services to help drive enterprise profitability.

As much as operational leaders recognize the need to gain new efficiencies and achieve digital transformation, they still may scratch their heads when they stumble upon ServiceNow. After all, this Enterprise Service Management (ESM) platform has its origins in IT, which leads to misperceptions that ServiceNow will be too complicated and non-intuitive for employees who aren't technology gurus. The reality is that ServiceNow was designed expressly to service the entire enterprise, with a user-friendly interface and powerful but simple analytics to drive intelligent decision-making and bottom-line efficiencies, regardless of the technical proficiency of the users. Let's explore four essential strategies for using the ServiceNow ESM model to serve maintenance and operations across the enterprise:

M&O teams can benefit from adopting a platform that was designed expressly to service the entire enterprise.

4.1 ALIGN REQUESTS WITH CORPORATE PRIORITIES AND GOALS:

Maintenance and operations has traditionally offered very personalized service, where each request is treated as a unique, custom task. While some requesters appreciate and benefit from this level of service, it is not scalable and cannot be systematically examined to identify and weed out inefficiencies (of which there are likely to be many). Consequently, for any organization focused on big-picture goals such as corporate sustainability or digital transformation, these manual processes represent a barrier to progress. ServiceNow helps organizations overcome these challenges by enabling all maintenance and operations requests to be logged into a centralized, web-based portal, which then tracks, prioritizes, and analyzes requests. From these prioritization and analysis processes, operational leaders are able to gain insights into how resources are being utilized, which, in turn, enables them to make more intelligent decisions as they align their resources with strategic priorities and goals.

ServiceNow analytics help M&O align their work to corporate priorities and goals.

4.2 CREATE ROBUST SELF-SERVICE FEATURES:

Maintenance, operations, and facilities workers typically expend great effort responding to requesters. Not only do they need to interface with the requester at the initiation of the service request, but requesters often want updates and routine status checks. Even more cumbersome, some requests are repetitive and redundant. Consequently, workers wastes precious time sorting out and responding to these issues. ServiceNow automates and streamlines management of service requests, with online, self-service facilities portals that provides 24/7 access to the status of work orders. Not only can employees submit requests automatically, but features such as floor-plan visualizations enable employees to self-verify whether an issue has already been reported. Furthermore, the ServiceNow-powered self-service portal offers built-in support for a full range of iOS and Android devices, ensuring employees can take full advantage of all self-service options.

MCKINSEY & COMPANY STUDY

Optimized plant maintenance, along with better management of spare parts and fuel, can increase earnings by

6.6%

Using enterprise resource planning to optimize operations at headquarters can lead to increased earnings by

1.3%

4.3 ASSIGN REQUESTS TO THE MOST QUALIFIED PERSONNEL:

Assigning service requests at most companies is based on tribal knowledge—the unwritten insights about who is most qualified for certain tasks, who prefers which tasks, and who is geographically suited to take care of certain tasks. All of this information that lives only in the brains of a select group of supervisors and managers should be codified into consistent best practices for assigning and managing work. ServiceNow offers a drag-and-drop feature for matching work tasks to the skills of the technician, as well as to geographic territory (i.e., in relation to current location) and to available inventory. This intuitive interface increases user adoption, and thus facilitates the centralization of institutional insights within a powerful, encompassing work management system.

ServiceNow offers intelligent case routing and robust self-service portals to reduce demand on M&O.

4.4 USE ANALYTICS TO MAXIMIZE RESOURCE UTILIZATION:

When humans assign work and assess resource allocations, they complete these tasks in an arbitrary way that is not necessarily systematic and optimized. ServiceNow performs these functions automatically using big-data analytics, ensuring volume and type of service requests can be taken into consideration in all resource allocation decisions. Furthermore, ServiceNow's dashboards display analyses of resource allocation metrics in simple terms, providing end-to-end visibility into work progress and enabling long-range planning. Occupancy decisions, for example, can be fully managed through ServiceNow. Operational leaders are able to run analyses of current and future occupancy needs, so they can reduce the need to reshuffle employees and reallocate work spaces.

4.5 CREATE INTERCONNECTIVITY WITH OTHER DEPARTMENTS:

Many of the work orders and requests fulfilled by maintenance and operations spill over into the territory of other departments. When new hires need an office to be set up for them, for example, they need more than just physical furniture and a computer; they also need IT to set up an internet connection and phone service, and HR to confirm which servers the employee will need access to. The interconnectivity among departments typically is facilitated through emails and phone calls, even though such processes are manual and non-standardized, and offer limited visibility into the real-time status of these processes. Moreover, these workflows are prone to delays and miscommunication that ultimately frustrate the requester and impede efficiency. ServiceNow offers an opportunity to get all departments that work on a single case—maintenance and operations, IT, customer service, finance, and so forth—to work from a single platform.

ServiceNow puts M&O on the same platform that other departments are already using.

Maintenance and operations leaders are continually looking for ways to improve delivery and management of core services. While it may not seem intuitive, ServiceNow is the optimal conduit through which these teams can rapidly resolve service issues, reduce errors and oversights, drive higher customer satisfaction, and ultimately drive digital transformation. This Enterprise Service Management (ESM) platform, in fact, was expressly developed with features and capabilities that help operational leaders align service requests to corporate priorities, create robust self-service options that reduce labor expenses, automatically route requests to the most qualified members of the team, use analytics to systematically maximize resource utilization, and create new lines of communication with other departments.

12%

admit they currently have no digital transformation strategy at all.

45%

are pursuing a formal digital transformation strategy.

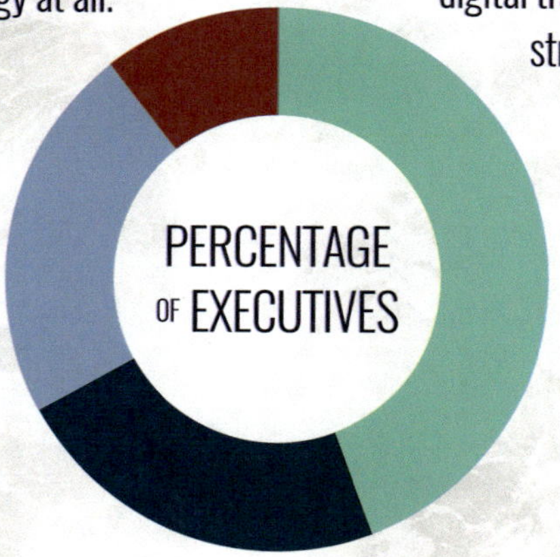

20%

are still in the planning stage.

23%

have started on a siloed approach, but lack an overarching plan.

451 RESEARCH STUDY, ENTERPRISE DIGITAL TRANSFORMATION STRATEGIES: TURNING DISRUPTION INTO DIFFERENTIATION

PILLAR 5

Managing Assets and SLAs Through ESM

MEET MARIA

Maria is the new vice president of operations for a fast-growing manufacturing startup. She was hired because of her reputation for bringing discipline and structure to unwieldy, untamed operational processes. Maria credits her success to an eagerness to embrace innovative technology solutions that address common pain points. As Maria examines the huge number of physical assets being acquired by her company, and the numerous Service Level Agreements (SLAs) being implemented, she notices that there are no standardized, centralized processes in place to manage these assets and SLAs over their full lifecycle. In fact, she finds a hodge-podge of non-standardized solutions, ad-hoc workflows, and non-centralized strategic planning initiatives. In talking with her colleagues in IT, Maria learns that computer assets and software SLAs acquired by her company are managed using ServiceNow. Never one to be intimidated by unfamiliar technologies, Maria thoroughly researches how the IT department uses ServiceNow to methodically review and track assets and SLAs, and to systematically weed out waste and inefficiency. Maria is wowed by not only the sophistication and dynamism of ServiceNow, but also its elegance and intuitive simplicity. She realizes that the ServiceNow platform could be used to manage all major physical assets, SLAs, and capital expenditures across the organization, from software licenses and network infrastructure to industrial equipment and vehicle fleets. In particular, she sees the potential of ServiceNow improve asset lifecycle management and to extract more value out of all SLAs.

ServiceNow will help you streamline asset management and maximize the value of your SLAs.

When operational leaders like Maria pursue ServiceNow as a tool for managing assets and SLAs, they are looking to dramatically streamline operations, reduce costs, improve customer satisfaction, and support digital transformation. A 2016 study by the research firm Gartner found that software asset management tools like ServiceNow enable organizations to reduce software expenses by a whopping 30%—and that these efficiencies are realized within the tool's first year of operation. What is the secret to success rates like this? Gartner says that software asset management tools like ServiceNow are particularly effective because they help organizations optimize software configurations and identify unused, underused, and misused software.

As an ESM platform, ServiceNow is primed to deliver on all of these promises. ServiceNow's streamlined, powerful design extends the reach of ESM far beyond IT applications to encompass all kinds of assets and SLAs. Indeed, ServiceNow has become invaluable to operational leaders who are increasingly being called upon to reduce costs of capital equipment and computer assets, and to extract more value out of SLAs. And for those in the business of providing service through SLAs, ServiceNow excels at streamlining the workflows necessary to comply with every customer's SLAs. Let's explore how to optimally take advantage of ServiceNow to manage assets and SLAs—and to ultimately support digital transformation across the organization:

Cost reductions are the main benefit of using ServiceNow for asset and SLA management.

> Software asset management tools like ServiceNow enable organizations to reduce software expenses by a whopping 30%—and these efficiencies are realized within the tool's first year of operation.
>
> 2016 GARTNER RESEARCH FIRM STUDY

5.1 CREATE A SINGLE SOURCE OF TRUTH:

Before organizations can make long-term resource planning decisions, they typically run reports on the status of physical assets and/or analyze service delivery against SLAs. Traditionally, these analyses have come from multiple sources that aren't necessarily up to date or complete, and that aren't necessarily shared with other departments even after they're prepared. As a result, precious time is wasted generating business insights that are often inaccurate, unreliable, and duplicative of prior efforts. ServiceNow can and should be configured to get every department working from the same system. By linking asset repositories with the ServiceNow Configuration Management Database (CMDB), a single system of action can be developed, reducing manual entry (and re-entry) of data and paving the way to track trends with KPIs (key performance indicators). Furthermore, information about software and hardware configurations can be automatically gathered, analyzed, and displayed on user-friendly dashboards. These insights facilitate intelligent, accurate decision-making on crucial operational issues like capacity, asset refresh, vendors, and so forth.

5.2 AUTOMATICALLY IDENTIFY UNDERUTILIZED, OVERUTILIZED, AND NONCOMPLIANT ASSETS:

As any operations employee will tell you, there's waste and inefficiency hiding in every corner of an organization—but there's not necessarily a cost-effective way to weed it out, at least through not manual means. Thanks to automation, ServiceNow is changing this paradigm. ServiceNow is designed to keep track of usage information and usage entitlement information for every asset that is entered into it, including cost information and data that represent the business value derived from the asset. These insights enable ServiceNow to automatically flag underutilized and overutilized assets, enabling companies to focus on fixing those issues—rather than spending precious time simply working to identify them. In other words, ServiceNow frees you up to focus on cost-savings measures like consolidating vendors, streamlining the types and models of assets you maintain, and canceling contracts that are duplicative and/or don't add value. Furthermore, when there's a compliance problem with an asset, it can be identified and rectified immediately, mitigating risk to the organization.

ServiceNow enables you to consolidate vendors and eliminate duplicative assets and contracts.

5.3 CENTRALIZE CONTROL OVER ASSET ACQUISITION AND DISTRIBUTION:

For departments to be able to acquire the assets and SLAs they want, many department leaders have traditionally eschewed centralized purchasing processes and gravitated to whatever lines of personal communication and workarounds they find most effective. These ad-hoc acquisition and distribution processes may work well for their individual beneficiaries, but they don't advance overall organizational efficiency. In fact, in the absence of centralized controls, the organization suffers from inadequate oversight and poor governance structures. ServiceNow enables operational leaders to impose strong, centralized controls over asset acquisition and distribution. Using a role-based service catalog, asset prices can be plainly displayed and automatically charged to the requesting department—assuming the requestor has authorization to do so. With centralization also comes the ability to perform routine compliance audits, enforce policies and organizational priorities, and identify and block the purchase of redundant assets.

5.4 RANK AND PRIORITIZE TO ACHIEVE SLA COMPLIANCE:

If you're in the business of servicing your customers' SLAs, you are intimately aware of how complicated and challenging SLA compliance management can be. Your SLA service team is made up of individuals with varying skills and strengths who ideally should be matched to specific types of jobs. Sometimes these folks travel across large geographies, or make contact with customers in radically different time zones. SLAs also often require highly specialized parts and components that must arrive in the right quantities at the right place and time. Traditionally, SLA service teams have juggled all of these moving parts manually, and used best professional judgment to set priorities for which services they fulfill in what order. ServiceNow automates all of this juggling and prioritization, so SLA compliance is optimized and cases don't slip through the cracks. Additionally, with all of this prioritization data in the ServiceNow system, you're able to run analytics to generate key metrics of service delivery, including trends in case volumes and types, employee satisfaction, and aged cases.

Companies that deliver services through SLAs can use ServiceNow to optimize how they deploy limited resources.

5.5 PROACTIVELY ENGAGE AS ASSETS AND SLAS APPROACH END OF LIFE:

Whether your customers are fellow employees or external clients and vendors, they all are using assets and SLAs that have defined lifecycles. In a best-case scenario, the average organization might have some manual processes in place to proactively alert certain customers and proactively work toward replacing assets and renewing SLAs. But these processes, if they exist at all, aren't systematic and consistent. ServiceNow offers an automated, fully integrated solution. With ServiceNow, you'll be able to eliminate reactive maintenance, prevent business disruptions, and optimize asset and SLA utilization across the full lifecycle. In this way, lapses in coverage can become a thing of the past, and sales and marketing teams can strategically make contact with their customers well in advance of lapses. Finally, with all of this information contained in a single centralized system, operational leaders will gain end-to-end visibility over the lifecycle of all assets and SLAs, including the ability to more easily run audits.

In the fast-paced world of asset and SLA management, operational leaders increasingly are being called upon to streamline and cut costs. To manage this process with consistency and rigor, these leaders need a centralized, encompassing system to automate and organize all of this data. ServiceNow delivers the digital transformation that operational leaders crave. The platform creates a single source of truth from which to track assets and SLAs, automatically flags assets that aren't being optimally utilized, centralizes control over asset acquisition and distribution processes, ranks and prioritizes SLA-related service obligations, and proactively engages with customers prior to the end of product lifecycles.

PILLAR 6

Planning and Executing Marketing Strategy Using ESM

MEET CLARK

Clark is the senior vice president of marketing at an established tech company. Clark's company has not been meeting its sales and growth targets, which has prompted every department to reexamine how to streamline its operations and boost customer retention and satisfaction. Clark sees opportunities for the marketing team to work smarter and more efficiently. He'd like to see the team use next-generation technology to automate and simplify day-to-day work management functions. He also wants the marketing team to roll out strategic initiatives that better align with long-term organizational goals. Clark's marketing team, however, isn't made up of tech gurus or early adopters. Clark recognizes that whatever changes he implements will need to be simple and intuitive, and team members on the ground will need to be able to see how new technology will benefit them in their daily work. Clark is aware that other departments in his organization have adopted ServiceNow, an Enterprise Service Management (ESM) system that has its origins in IT. But Clark is skeptical that ServiceNow could be useful for planning and executing marketing strategy, and streamlining and simplifying day-to-day work. What does Clark need to understand about ServiceNow to be convinced ServiceNow is right for his marketing team?

Companies around the world are grappling with how to utilize technology to generate big-data insights that will help them execute on long-term strategic initiatives of all kinds, including marketing. A 2016 study by PwC found that while business leaders understand the need to use big-data analytics as the foundation of decision-making, they aren't actually taking advantage of technology to incorporate data into their decision-making frameworks. Nearly two out of three executives reported in the PwC survey that decision-making in their organization is only rarely or somewhat data-driven.

Marketing, like every other department, should be using big-data insights to drive decision-making processes.

ServiceNow offers a path forward for departments that are seeking to make optimal use of big-data insights. With ServiceNow's centralized work management platform, departments don't just gain the critical business insights they need to develop and execute on strategic corporate goals and priorities; they're also immediately able to streamline and automate their workflows in support of enterprise-wide digital transformation. ServiceNow, which has its origins in IT, does not seem at first glance like a viable solution for a non-technical department like marketing. However, the IT gurus who created ServiceNow intuitively understood that all enterprise work—regardless of department—can and should be managed through an enterprise work management platform. Thus, ServiceNow was designed to be simple and intuitive to use, with powerful analytics that can drive intelligent decision-making for marketing and other transformation-minded departments across the organization. Let's explore how you can optimally use ServiceNow to plan and execute marketing strategy that will advance your organization's long-term business objectives and support digital transformation:

ServiceNow can help marketing departments plan and execute on strategies that are aligned to enterprise-wide goals.

6.1 ALIGN PROJECTS TO STRATEGIC BUSINESS GOALS:

Marketing departments routinely spend precious resources on tasks that are not aligned with the organization's strategic interests and goals. The challenge is that organizations often overlook the important task of creating clearly defined goals for marketing, making it impossible for marketing leaders to properly evaluate which projects they should pursue to advance enterprise-wide goals and priorities. ServiceNow makes it easy for marketing leaders to align all projects to business strategy, by generating a broad spectrum of KPIs that illuminate the relative value of various projects in relation to core business strategy. These powerful analytical capabilities will eliminate subjective decision-making and help marketing leaders allocate proper resources to initiatives that have the potential to truly transform the organization. Once your marketing team begins inputting project data into ServiceNow, you'll find that generating these analytical capabilities is intuitive, automated, and available to you in real time.

servicenow

ServiceNow generates a broad spectrum of KPIs to illuminate the relative value of various projects in relation to core business strategy.

6.2 BRING CONSISTENCY AND STRUCTURE TO MARKETING REQUESTS:

In most organizations, marketing department assignments and requests are transmitted through a variety of unstructured channels, including emails, phone calls, and department meetings. Marketing team members spend copious time sifting through these requests, trying to make sense of what they should work on and when. ServiceNow puts an end to this lack of structure by imposing standards and templates on work processes. In this way, marketing professionals are able to focus on what they do best—creative services, content development, marketing campaigns, and so forth. Additionally, the marketing department is able to create its own cloud-based "storefront" for receiving requests. With all work properly structured in one centralized location, marketing department leaders also are able to pull up comprehensive, real-time dashboards that track status, resources, and demand.

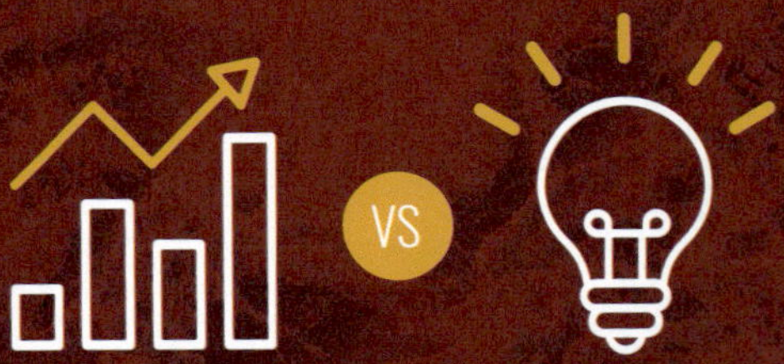

> While business leaders understand the need to use big-data analytics as the foundation of decision-making, they aren't actually taking advantage of technology to incorporate data into their decision-making frameworks.
>
> 2016 PwC STUDY

6.3 ROUTE REQUESTS AUTOMATICALLY TO THE RIGHT TEAM MEMBERS:

The IT wizards who developed ServiceNow understood that not every employee is created in the same mold, and that different employees have unique strengths and skill sets for solving specific types of issues. Thus, ServiceNow is designed to facilitate automated routing of service requests to the individuals who are best equipped to resolve them. In the same way that ServiceNow has introduced automated routing of IT tickets to specific agents, so too can ServiceNow be used by marketing departments. In fact, requests for marketing collateral can be assigned automatically within ServiceNow to the marketing team members best equipped to handle them; this is done automatically based on how the request has been categorized. This approach not only saves on time and labor, but also brings consistency and standardization to how team members are assigned certain types of tasks. Marketing teams that can reliably complete assignments in a predictable, consistent manner will inevitably end up with predictable, consistently high quality across every project they tackle.

6.4 USE SELF-SERVICE PORTALS TO SHARE MARKETING WISDOM AND INSIGHTS:

Marketing teams work hard to create polished customer campaigns and develop marketing strategy aligned with long-term business goals. While you retain full control over what your team works on, other departments are simultaneously developing their own customer-facing products that you don't control, such as customer service materials, sales sheets, and annual reports. You want the creators of these products to understand the importance of brand consistency and cohesion, even though you're not in a position to personally connect to everyone in your organization to share these wisdom and insights. In fact, you may not even find out about some of these products until after they're completed. ServiceNow provides out-of-the-box infrastructure to easily create self-service knowledge portals focused around key marketing concepts and insights. These portals are easy to set up, easy to maintain, and easy for others to find and read. On them, you'll be able to share wisdom, insights, tips, and best-practices recommendations that can dramatically enhance brand consistency and cohesion.

ServiceNow uses intelligent case routing and self-service portals to help marketing departments make the most of limited resources.

6.5 ELIMINATE REACTIVE WORK AND ACCURATELY PREDICT DEMAND:

Many marketing team leaders plan about one day or one week in advance. Although they intend to engage in long-range planning, they often become consumed by day-to-day demands. Consequently, they're challenged to truly seize the moment and take a proactive approach to managing demand. Over time, they involuntarily fall into an unfortunate, frustrating pattern of allowing other departments to dictate their projects. This quickly turns them into a team that thinks reactively rather than proactively. ServiceNow allows marketing departments to organize all of their workflows, resources, and assignments, so marketing leaders can visualize and predict demand, and then make informed, strategic decisions. When marketing departments are able to intelligently predict demand and proactively adjust resources and plans, they won't let anyone else dictate their work—they'll know what's coming and how to plan to meet this demand. And their efficiency and productivity go up as a result, along with their ability to keep the brand one step ahead of the competition at all times.

There's no reason to think that ServiceNow will be too complicated and non-intuitive for your marketing team to use. ServiceNow was designed to service and drive digital transformation across the entire enterprise, regardless of the technical proficiency of end users. Your marketing team will get maximum benefit from ServiceNow by using this powerful ESM platform to align projects to strategic business goals, to bring consistency and structure to requests that come into marketing, to route requests automatically to the optimal team members, to share marketing wisdom and insights through self-service knowledge portals, and to proactively anticipate demand and plan accordingly.

2 OUT OF 3

executives reported that decision-making in their organization is only rarely or somewhat data-driven.

2016 PwC STUDY

PILLAR 7

Enhancing Financial and Budget Management with ESM

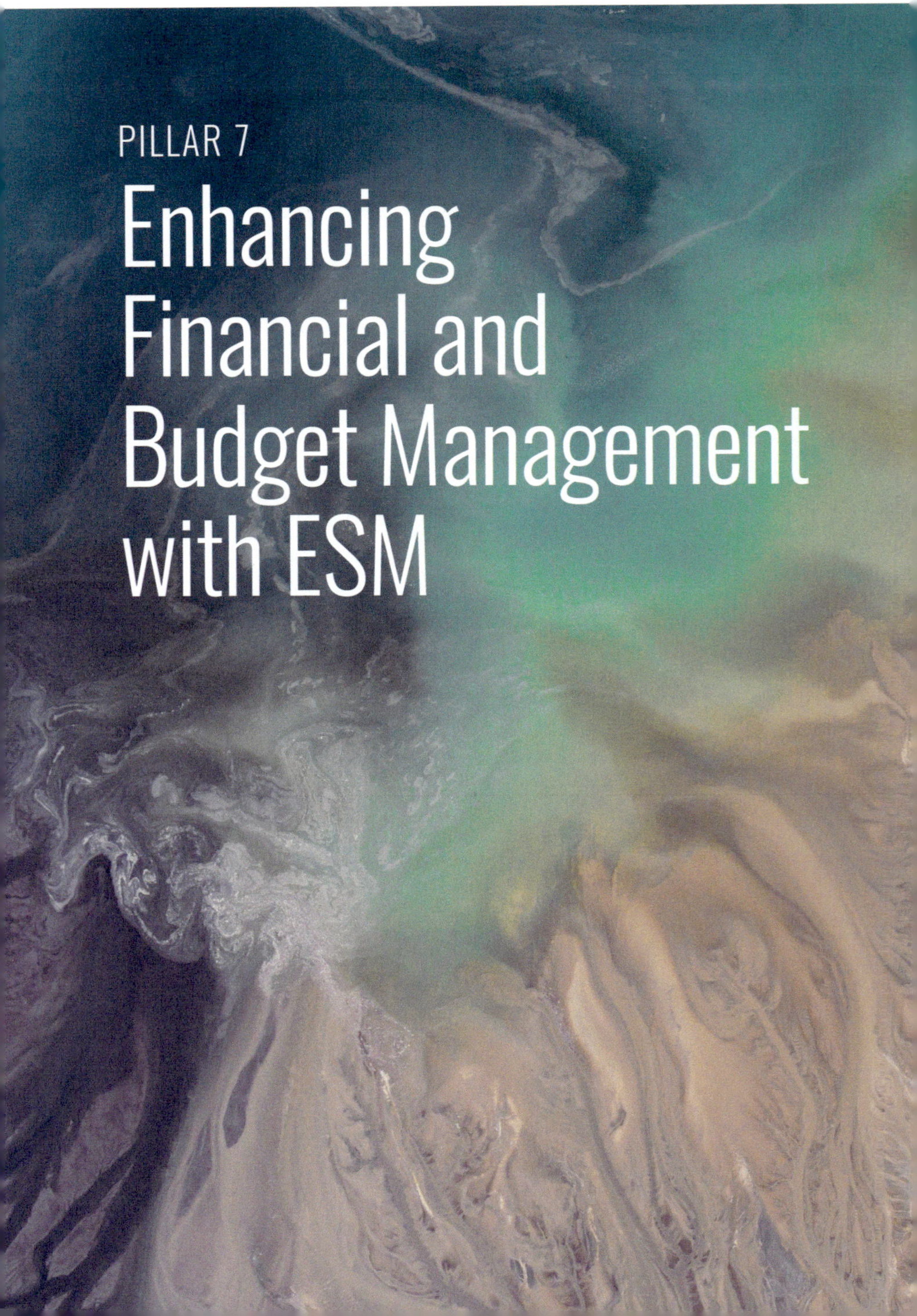

MEET CLARK

Rebecca is the newly installed chief financial officer for a Fortune 500 company. Being an external hire, Rebecca is shocked to discover outdated and inefficient budgeting practices in the finance department. Records are maintained on spreadsheets that are individually compiled and linked, and then manually reconciled. She readily finds embarrassing mistakes, wasted resources, and general lack of transparency and efficiency. Although Rebecca knows that she'll be able to choose from an array of modern, cloud-based financial management solutions to replace these antiquated budgeting processes, she isn't sold on any of them. In fact, she is concerned about rolling out yet another layer of technology that will only be directly beneficial to the financial department. Rebecca's preference is to more effectively integrate finance operations into the organization as a whole, and to create bridges of understanding and collaboration among departments. Financial management, after all, is part of an integrated business ecosystem that powers profitable organizations. Hence, over the long term, Rebecca wants every department to take more ownership of the budgeting process, and become more engaged with cost containment and long-term planning. This line of thinking leads Rebecca to pursue the most logical solution—ServiceNow—even though she knows it's going to raise a lot of eyebrows.

Finance departments should be integrating their budgeting processes into the organization as a whole.

The interest in ServiceNow among forward-thinking CFOs like Rebecca makes sense. CFOs are increasingly feeling pressure to integrate budgeting and financial operations with the overall strategic direction of their company. A 2016 survey conducted by the Argyle Executive Forum and Host Analytics found that more than 70% of CFOs believe that their relationship to the CEO has become more collaborative in recent years. The survey also found that 53% of CFOs believe that it's a "big priority" in their organization to help non-finance employees understand and engage with the organization's financial objectives. This increasing emphasis on collaboration and engagement with financial matters across the organization has made ServiceNow an especially attractive solution. If ServiceNow already is the go-to resource for IT departments—and ideally for operations, facilities, HR, and other departments—then it's a no-brainer to extend ServiceNow's reach to digitally transform financial planning as well.

Given that two out of three CFOs believe that the finance leader role will fundamentally change as traditional finance tasks are automated and managed through shared services centers, financial leaders are increasingly recognizing the need for digital transformation solutions like ServiceNow. ServiceNow is a cloud-based Enterprise Service Management (ESM) platform that is designed to encourage collaboration and synergy among employees across the organization. Although ServiceNow was originally developed by IT professionals to manage IT projects, ServiceNow has been greatly expanded in recent years to encompass financial management and business applications as well. Let's explore how you can use the ESM delivery model that emphasizes automation, efficiency, and collaboration to optimally enhance financial and budget management across your organization:

Using ServiceNow for your budgeting processes promotes enterprise-wide collaboration and synergy.

7.1 ALIGN BUDGETS TO BUSINESS REQUIREMENTS AND LONG-TERM STRATEGY:

Finance departments have traditionally focused on putting together budgets that can pass muster with shareholders, auditors, and the C-suite. What often gets short shrift is a focus on aligning the budgeting process to business strategy. Studies have shown that budgeting tends to be a short-term-oriented activity that disregards strategy and becomes a "scheming game." ServiceNow ensures that you can properly prioritize long-term business strategy during the budgeting process. ServiceNow's financial management applications offer organizational and visualization features that enable budgets to be intuitively aligned with long-term goals. Moreover, ServiceNow can support multiple budget models that encompass multiple hypothetical use cases, ensuring you can fully tease out all of the potential tradeoffs as you weigh short-term vs. long-term business needs.

7.2 MAKE BUDGET INFORMATION TRANSPARENT AND RELATABLE:

Financial departments communicate in complex spreadsheets that convey complex information, nuances, and caveats. Unfortunately, this presentation style isn't going to be readily understandable to non-financial employees. ServiceNow presents your budget information in transparent and relatable ways, so every department can understand and trust the budget that everyone must live with. ServiceNow's approach to financial management will require a major cultural shift for your finance team, but that's precisely the goal—it's all part of the digital transformation journey that finance is taking. You want to encourage employees outside of finance to engage with and buy into this process as well, and they are only going to do that if you make the budgeting process transparent and relatable to them. Once your audience is invested, you can tap into this heightened level of engagement and get them to rally around the big-picture financial issues. In other words, you want to turn them into protectors and defenders of the company's overall financial health, and appreciate the critical roles they individually play in influencing overall financial performance.

ServiceNow makes the budgeting process transparent and relatable to non-finance employees.

7.3 CREATE A CULTURE OF BUDGET ADHERENCE:

In a typical organization, every department has a fiefdom-like mentality, focused on getting as many resources as possible and secretly ignoring or undermining enterprise-wide priorities that run contrary to the department's own goals. Thus, the budgeting process represents one of the

53%

of CFOs believe that it's a "big priority" in their organization to help non-finance employees understand and engage with the organization's financial objectives.

2016 ARGYLE EXECUTIVE FORUM & HOST ANALYTICS SURVEY

few times that leaders from every department sit down at the same table and can be held accountable for any disconnects and discrepancies between their own priorities and those of the organization. ServiceNow ensures that these discrepancies are systematically illuminated and laid bare—not just during annual budget planning sessions, but also throughout the year. Especially because the costs of initiatives like digital transformation are high, it is paramount that every department has a way to visualize the alignment (or lack thereof) between requested vs. allocated vs. actual resources and spending. ServiceNow allows these relationships to be visualized at the project portfolio level. Additionally, ServiceNow's business capability maps and dynamic scoring engine will keep department leaders focused on how they can improve and optimize resource allocation for the entire organization, instead of for their individual departments. The end result is a culture of budget adherence across the organization. Individual departments will begin to recognize that budget discrepancies cannot be obfuscated and overlooked, and that no one can get away with financial decisions that run counter to agreed-upon priorities and goals for the organization.

7.4 MAKE COSTS OF CONSUMPTION EXPLICIT:

In the same way that departments tend to look the other way when budgeting priorities run contrary to their own, many departments—or at least many individuals within them—try to ignore and downplay the costs of their consumption. The root of this problem is the exceptional ability of humans to compartmentalize. In other words, we recognize that we make problematic spending decisions, and we separately recognize that problematic spending affects the financial health of the company, but we don't allow ourselves to recognize that it is our problematic spending decisions that are adversely affecting the company's financial health. When everyone compartmentalizes in this way, it can have enormous ripple effects on overall financial performance. The solution is to make the costs of individual consumption explicit for every employee. ServiceNow is primed to achieve this goal, offering organizations the ability to build a service catalog that documents and standardizes the cost of services across the organization. Every time you're thinking about charging another department for services, and every time you want to authorize a purchase order, ServiceNow can spotlight the costs of your decision. This will force every employee to think critically about every spending decision they make, in the context of whether it strategically aligns with enterprise-wide priorities and goals.

ServiceNow will ensure employees are fully aware of the financial implications of their spending decisions.

As finance teams are increasingly called upon to integrate budgeting processes into the organization as a whole, the need for an ESM solution like ServiceNow will grow. ServiceNow's financial management and business applications are designed to drive digital transformation, with simple and intuitive interfaces that require minimal learning curves and zero IT proficiency. You can use ServiceNow to optimally enhance financial and budget management by aligning your budgets to business requirements and long-term strategy, making budget data transparent and relatable to non-finance employees, creating a culture of budget adherence across the organization, and making the costs of consumption explicit and unavoidable as spending decisions are made.

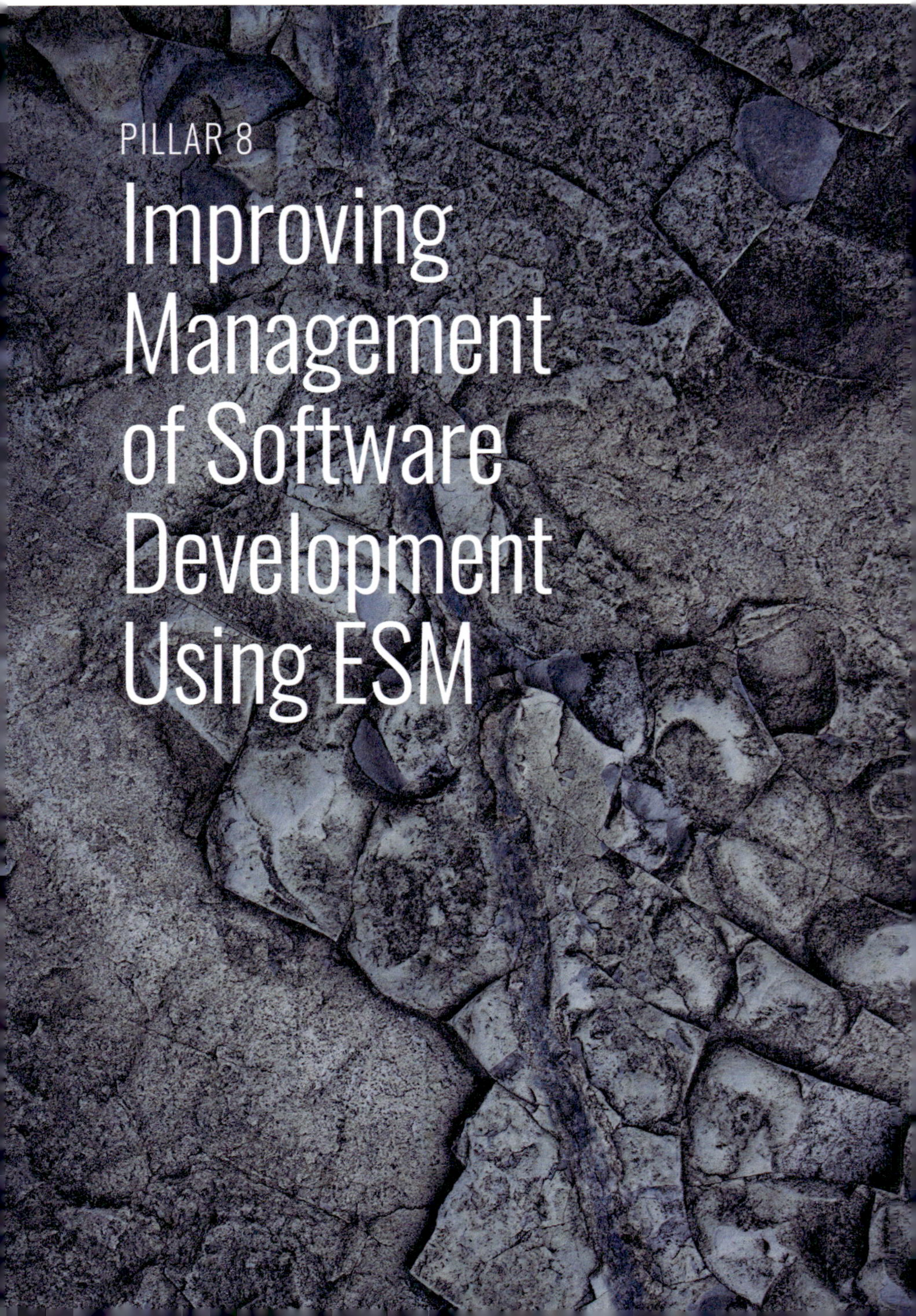

PILLAR 8

Improving Management of Software Development Using ESM

MEET MATT

Matt is the global vice president of technology development for an established manufacturing company. Among his responsibilities is overseeing the company's software development team, which has successfully developed a number of apps and technology upgrades across the enterprise. The problem is that from Matt's perspective, the development team functions as a black box. Matt cannot peer into the inner workings of this department and do his job. In other words, he cannot guide and manage the work and priorities of this team as effectively as he needs to. During biweekly status meetings with development team leaders, Matt must essentially take these managers at their word and make decisions based on limited data. He can't, for example, log into JIRA and independently extract business insights on the progress of an enterprise app under development. Consequently, he doesn't know exactly what the development team is working on at any given time, can't investigate and weed out inefficiencies and redundancies, and can't assess adherence to deadlines, budgets, and other project specifications. These shortcomings lead Matt to search for another system that can deliver the business insights he needs, while also integrating seamlessly with JIRA and the other native platforms the development team already is using. ServiceNow, an enterprise work management platform already being used by the company's IT service management team, emerges as Matt's leading choice.

servicenow

ServiceNow enables project and operational leaders to work on the same platform as software developers.

By 2022, 80% of revenue growth will depend on digital offerings and operations.

IDC PREDICTION

When enterprise leaders like Matt are in the dark searching for digital transformation solutions, they often seize upon ServiceNow as an approach for increasing visibility into work processes of all kinds. Just as project managers need real-time data on a project's status to keep it on track, Enterprise Service Management (ESM) leaders need a global way to evaluate performance of projects, and guide teams toward greater productivity and efficiencies. A recent survey of project managers found that when relevant and up-to-date information about projects is routinely made available, the likelihood of project success increases by 48%. ServiceNow's agile development applications have been specifically designed to offer this enhanced level of visibility to people who aren't software developers. And ServiceNow does these things without creating new layers of bureaucracy, and without setting the stage for ongoing, burdensome maintenance responsibilities.

When ServiceNow is rolled out in a development environment, its main function is to connect all of the people who are following the software development lifecycle. Once this data is shared and analyzed by the ServiceNow engine, project portfolio managers, operational leaders, and even the C-suite gain key business insights into the status of development projects. Then, from within ServiceNow, they can control and direct the work of the development team, manage iterative enhancements and fixes, and plan for the product's rollout and adoption phases. Let's explore how to optimally use Servicenow's ESM capabilities to improve management of the development process from inception to deployment—and ultimately to drive digital transformation:

Businesses can keep tabs on deadlines, budgets, and other specs for software development projects from within the ServiceNow platform.

8.1 DON'T USE SERVICENOW TO REPLACE EXISTING DEVELOPMENT PLATFORMS:

If you tried to replace any of your development team's native platforms with ServiceNow, you'd have a revolt on your hands. Fortunately, it is a bad idea anyway. When you add ServiceNow to your software development workflows, it should be in addition to—not in place of—your existing systems. ServiceNow can be configured to integrate seamlessly with development systems like JIRA, enabling data from the native system to be transferred into ServiceNow continuously and in real time. Then, you want to configure ServiceNow to automatically analyze all of this data and serve it up as business intelligence that is readily understood by your non-developer audience. In this way, ServiceNow becomes the single system of record through which all relevant departments collaborate to fully manage the development lifecycle.

> **ServiceNow integrates seamlessly with native development platforms like JIRA.**

8.2 INCREASE PLANNING ACCURACY WITH BUILT-IN SERVICENOW FEATURES:

When you're trying to plan for the rollout of a new software product, you're in constant communication with your development team. You need to manually obtain status updates, timelines, delivery estimates, and so forth, and then you need to convert this raw information into progress reports, charts, and other visualizations. In the end, the accuracy of your planning efforts is likely to be uneven at best, and woefully off base at worst. With ServiceNow, there's no need to do any of these steps manually. ServiceNow comes preloaded with features that allow you to easily create burn-down charts, progress and release-planning boards, and team velocity and story-point estimations. Using relational data in ServiceNow, you'll be able to map defects, enhancements, incidents, and problems. You'll also be able to use role-based access to control who can view and/or edit these records, and trace development changes back to the originating request or issue.

8.3 TAKE ADVANTAGE OF SERVICENOW'S CONSUMER-GRADE TOOLS:

Although ServiceNow isn't intended to replace any of your development team's native platforms, you might be surprised when your development team turns to ServiceNow on its own. That's because ServiceNow's agile development capabilities include a variety of easy-to-use, consumer-grade tools that can significantly reduce the length of the development lifecycle. ServiceNow templates, scripts, and integrations can be used in tandem with universal components, platform services, APIs, and mainstream web technologies to inject efficiency and elegance into every development project. While you shouldn't force these tools on your development team, there's no reason you can't educate your team on the potential of these tools, as well as encourage your team to take advantage of these tools whenever these tools have the potential to enhance quality and speed of delivery.

> ServiceNow comes pre-loaded with consumer-grade development tools to enhance the work of developers.

8.4 CONNECT DEVELOPMENT FEATURES TO BUSINESS DRIVERS:

As you use ServiceNow to manage the development process for new software applications, you want to ensure that the features of these products are optimally aligned with strategic priorities and goals. ServiceNow is

Digital transformation initiatives are currently led by different individuals:

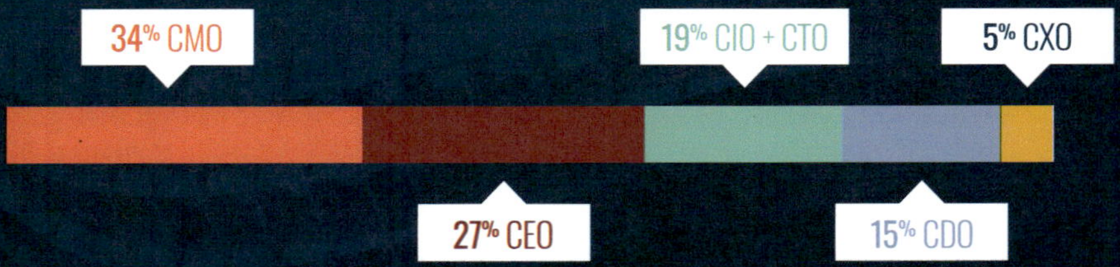

STATE OF DIGITAL TRANSFORMATION REPORT, ALTIMETER GROUP, 2016

designed to toggle seamlessly between managing service processes and development processes. That means ServiceNow can display the alignment (or lack thereof) between the work of your development team and business drivers, helping you to better guide and direct the day-to-day work of your development team. You want to get to a place where every feature and capability that your developers build is driven by business needs. This will create the alignment you need to get the most out of all of your development projects.

8.5 INCORPORATE WATERFALL AND HYBRID DEVELOPMENT METHODOLOGIES WHERE APPROPRIATE:

Agile development methodologies have been pushed hard in recent years as the gold-standard, best-practices approach for software development work. With agile methodologies, groups work in short cycles toward the final product, often dealing with rapidly changing specifications along the way. While Scrum and other agile methodologies have become dominant, they aren't always appropriate or ideal. ServiceNow recognizes this reality, offering maximum flexibility for managing development projects. ServiceNow gives you the ability to manage agile products alongside waterfall projects, as well as projects that combine elements of both agile and waterfall methodologies. You can even add waterfall elements to projects that were originally initiated using the agile methodology. As a leader, you want to encourage and promote this flexibility to ensure the best possible development outcomes. You can count on ServiceNow to rise to the occasion to meet these needs.

ServiceNow offers the visibility and business intelligence that organizations crave to properly manage the work of their software development teams. Rather than replace any of the native platforms that developers use, ServiceNow integrates with these platforms, giving non-developers an elegant intuitive interface to track and direct the work of the development team. You should take advantage of built-in ServiceNow tools to increase planning accuracy, as well as to boost quality and speed of delivery.

You also want to align development features with business needs and priorities, and incorporate waterfall and hybrid development methodologies where appropriate. From a software development perspective, these are the hallmarks of digital transformation.

> Research reveals that the companies that do best follow bold and disruptive strategies. They make big bets on new technologies and business models, champion a test-and-learn culture where every failure is an opportunity to improve, and launch change programs that transform their whole business.
>
> McKINSEY RESEARCH

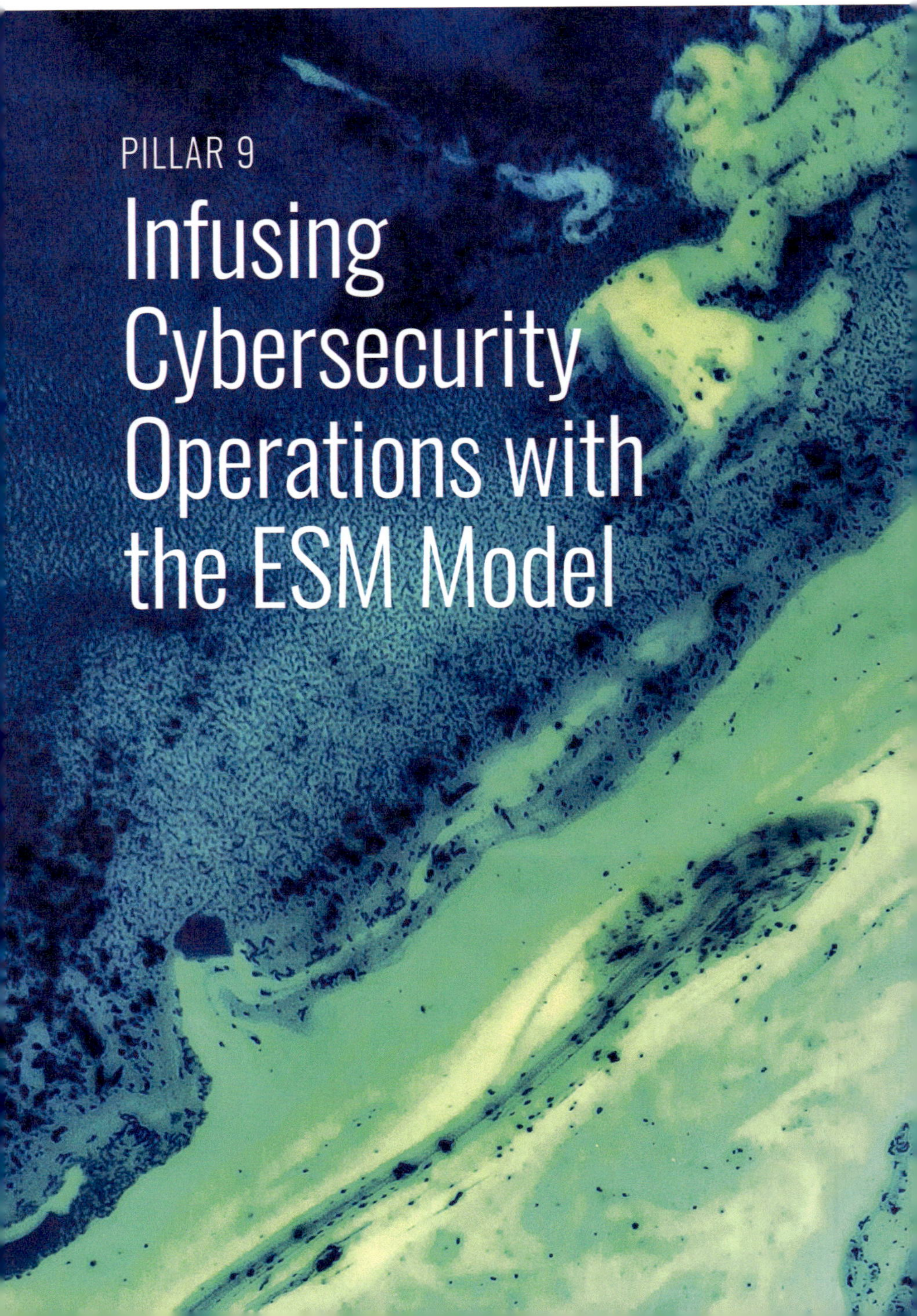

PILLAR 9

Infusing Cybersecurity Operations with the ESM Model

MEET JUAN

Juan is the long-time vice president of global security for a multinational firm that is growing increasingly concerned about its cybersecurity vulnerabilities. Juan is tasked with undertaking a comprehensive review of all of the security products, services, and resources across the company, as well as completing an inventory of all of his company's SLAs with security vendors. From this effort, he hopes to eliminate redundancies, direct more resources to the most mission-critical areas, and improve incident response times. As Juan embarks on this comprehensive review, he realizes that his team isn't tracking the metrics that would reveal how the organization's cybersecurity operations are performing. Juan cannot answer questions like: Is the global enterprise as secure as it could be? Is the company getting better or worse at responding to incidents? Are top-priority issues slipping through the cracks because incident responders are overloaded? As Juan chats with his counterparts in IT, he learns that ServiceNow offers enterprise-level security response capabilities that manage the process of assessing and responding to vulnerabilities, as well as that provide crucial business context to optimally prioritize and remediate incidents. Juan is especially impressed that, on a fundamental level, ServiceNow will connect IT to security operations; all of these vitally important connections are forged in a manual, one-off manner right now, which Juan realizes cannot possibly be efficient or best practices.

servicenow

ServiceNow provides cybersecurity managers with critical business context for optimally prioritizing and remediating incidents.

The deficiencies in cybersecurity operations that Juan has picked up on are endemic across organizations of all sizes and industries. In a 2016 survey by the Enterprise Strategy Group, nearly three in four cybersecurity professionals reported that incident response in their organization tends to be based on informal, manual processes. Furthermore, the survey found that one in three organizations spends at least half of all incident response time on manual processes, leading to inefficiencies and delays. The root of this problem lies with the fact that too many organizations have not defined and formalized a standardized, best-practices set of incident response workflows. Thus, bottlenecks, inconsistencies, and lethargic response times become par for the course.

ServiceNow is designed to eliminate these informal, manual workflows by introducing automation, collaboration, and intelligent workflows—the hallmarks of digital transformation via the ServiceNow Enterprise Service Management (ESM) philosophy. With ServiceNow, you'll gain greater visibility into security operations, as well as dramatically shortened discovery and remediation times. Indeed, in many cases, response times for top-priority incidents can be cut from days to a mere minutes. Let's explore how to optimally infuse the ESM model into cybersecurity operations to better protect the organization:

Response times for top-priority incidents can be cut from days to a mere minutes using ServiceNow.

9.1 MAP THREATS TO BUSINESS INFRASTRUCTURE AND SERVICES:

When most cybersecurity teams attempt to prioritize threats, they do so in a silo, without fully understanding the business units whose assets, services, and infrastructure they are tasked with protecting. Consequently, the way that cybersecurity teams prioritize threats is often misaligned with business priorities and needs. ServiceNow eliminates this challenge by automatically mapping threats to business assets, services, and infrastructure. In fact, the maps show all dependencies of affected systems, so security teams can get a comprehensive overview of all vulnerabilities affecting a given system. When cybersecurity analysts are able to visualize what they're protecting and see the interdependencies among assets, they bring a higher level of focus and resolve to their threat prioritization activities. These mapping visualizations also help analysts devise strategies for minimizing downtimes and managing change requests.

Nearly 3 in 4 cybersecurity professionals reported that incident response in their organization tends to be based on informal, manual processes.

2016 ENTERPRISE STRATEGY GROUP SURVEY

9.2 TAKE ADVANTAGE OF BEST-PRACTICES INCIDENT-RESPONSE WORKFLOWS:

One of ServiceNow's greatest strengths in the security field is its ability to immediately get your cybersecurity team to follow articulated, best-practices workflows for incident response. Given that many organizations still struggle with these processes, you want to take full advantage of these powerful, built-in workflows. Here's just one example: When a critical vulnerability is identified, ServiceNow can take care of automatically initiating an emergency patch approval request, so the analyst doesn't need to manually generate it. Then, once the patch has been approved and applied, ServiceNow auto-triggers an additional vulnerability scan to verify the issue is resolved. On the flip side, when a non-critical vulnerability is identified, the security analyst can click a single button in ServiceNow to generate a change request that goes to IT; this workflow allows the analyst to immediately return his or her attention to managing higher-priority threats. Spending as little time as possible on these simple, basic tasks is essential for any cybersecurity analyst—at least if they want to stand a fighting chance against increasingly sophisticated cyberattacks.

servicenow

ServiceNow enables cybersecurity teams to establish articulated, best-practices workflows for incident response.

9.3 ENRICH INCIDENTS WITH ALL RELEVANT THREAT INTELLIGENCE DATA:

When a high-priority vulnerability is identified, cybersecurity teams typically spend copious time manually researching the vulnerability, including searching for knowledge-base articles, manuals, and other insights on the topic. Not only is this process inefficient, but it also eats up precious time. ServiceNow enables your cybersecurity team to automatically retrieve threat intelligence and attach it to any incident. In fact, ServiceNow can be configured to pull intelligence data and articles from its own security knowledge base, from third-party threat feeds, from STIX and TAXII, and from any additional custom feeds you choose. Additionally, even before the analyst opens the incident file, ServiceNow can take care of performing targeted malware scans and pulling running processes from an affected endpoint. You want to take full advantage of these automated research features to speed up your cybersecurity team's ability to respond optimally to cyber vulnerabilities.

9.4 MONITOR SLAs TO IMPROVE MISSION-CRITICAL ACCOUNTABILITY:

A comprehensive cybersecurity program is only as strong as its weakest link, and unfortunately, there can be many weak links in your cybersecurity systems. Robust cybersecurity programs are made up of many moving parts that are controlled by both external vendors and internal team members. Thus, for cybersecurity responses to be effective, each of these players must do their jobs on time and to exacting specifications. ServiceNow has been designed to allow organizations to automatically and continuously monitor SLAs (service-level agreements) for non-compliance. When tasks aren't completed on time and in accordance with SLA thresholds, ServiceNow can send reminders to the responsible party and/or escalate the tasks. These monitoring processes inject an important layer of accountability into SLA management, and ensure that mission-critical security tasks never slip through the cracks.

ServiceNow can be configured to automatically and continuously monitor cybersecurity SLAs for noncompliance.

9.5 USE METRICS AND POST-INCIDENT REVIEWS TO DRIVE CONTINUOUS IMPROVEMENT:

Cybersecurity teams are typically so overwhelmed putting out fires that they don't properly debrief after incidents and report out on incident-response performance. Part of the reason for overlooking this important step is that cybersecurity teams traditionally have not had access to simple, standardized systems for generating and analyzing post-incident reviews; they are forced to rely on decentralized spreadsheets and email. In contrast, once an incident is closed in ServiceNow, ServiceNow automatically generates and distributes an assessment to the entire team, including a time-stamped post-incident review that serves as a historical audit record. These assessments include a number of KPIs that are built into ServiceNow to track key performance metrics; users also can create their own custom KPIs. You want to take full advantage of these features to drive continuous improvement.

Cybersecurity teams no longer need to rely on manual, informal processes for responding to vulnerabilities and threats. ServiceNow's powerful automations and elegant workflows will bring standardization and rigor to incident response, and increase visibility over mission-critical processes and tasks. To optimally infuse ServiceNow's ESM delivery model into your cybersecurity operations and drive digital transformation, you want to map threats to business infrastructure and services, take advantage of ServiceNow's best-practices workflows for incident response, enrich incident files with relevant threat intelligence data, monitor SLAs to improve accountability, and use metrics and post-incident reviews to drive continuous improvement.

> Across the globe, businesses are racing to implement new technologies, using data to innovate and grow in an increasingly interconnected world. But as companies become more reliant on cyber capabilities, they must also recognize and manage the associated risks. To compete in the emerging digital world, companies must protect themselves from cyber risks, and become more resilient to cyber shocks—large-scale events with cascading disruptive consequences.
>
> 2018 PwC SURVEY, GLOBAL STATE OF INFORMATION SECURITY

PILLAR 10

Integrating and Coordinating ESM Across Business Functions

MEET SYDNEY

Sydney is the chief operating officer for a manufacturing company committed to a wholesale digital transformation. The C-suite has bought into an ambitious plan to roll out ServiceNow in phases to departments across the enterprise, from HR to sales to security. Sydney is tasked with working with the company's vice presidents to implement this ambitious, three-year plan. While Sydney has a clear path forward, she isn't sure what to expect over the long term. Naysayers already are questioning the organization's ability to withstand such a dramatic and fundamental shift in how work is performed and managed. Sydney reaches out to her senior vice president of technology to gain insights on ServiceNow, as the IT department already has been using ServiceNow for nearly a half a decade. He encourages Sydney to do as much preparation and research as possible, especially given her all-encompassing vision for ServiceNow. She recognizes the challenge ahead and starts soliciting advice and guidance from those who have already gone down this path.

The initial legwork that goes into Sydney's ServiceNow implementation journey is one of the important but overlooked steps in positioning a company for long-term success with ServiceNow. Companies often struggle to use powerful technology like ServiceNow to drive fundamental workplace change, especially when the company is seeking more than a siloed, department-by-department approach to digital transformation. Indeed, well-coordinated planning is essential to convince all stakeholders of the value and importance of enterprise-wide transformation. A study by the MIT Sloan Management Review and Capgemini Consulting found that 63% of managers believe the pace of digital transformation is too slow in their organization, primarily due to a perceived lack of urgency and poor communication about the benefits of upgrading.

Enterprise-wide digital transformation depends on keeping all stakeholders engaged and on the same page.

For a ServiceNow-fueled digital transformation to find enterprise-wide success, no department can afford to get stuck behind a dreaded "transformation wall," in which the department become unable to break through and realize its full potential. Every department must buy into the coming changes fully and without reservations. ServiceNow's key selling point is the opportunities it creates to connect workflows and business processes across your organization. Departments that were traditionally siloed will suddenly be able to share key business intelligence, and collaborate on initiatives that drive efficiencies and profitability across the entire organization. As you work to build this interconnected business ecosystem, it's crucial to be aware of what you should—and shouldn't—do to optimally position each department in your organization for success. Let's explore five essential strategies for integrating and coordinating Enterprise Service Management (ESM) across your business functions and departments:

10.1 REQUIRE EVERY DEPARTMENT TO LINK PLANS TO ENTERPRISE-WIDE PRIORITIES:

After the initial ServiceNow kickoff meeting has passed and your department heads return to their offices, they are going to quickly lose sight of what the organization is trying to achieve with ServiceNow and how the master plan aligns to their individual digital transformation initiatives. Consequently, many of them will develop implementation plans for their department that reflect their own wants and needs, but that won't be aligned to enterprise-wide goals. This is an unfortunate outcome, because to get the most out of ServiceNow, each department must use ServiceNow in a way that supports and furthers the objectives of the organization as a whole. The solution to this challenge is to require every department to link their ServiceNow implementation plans to the original objectives articulated during the kickoff meeting. When they're challenged to articulate these linkages as part of each implementation planning presentation and report they prepare, they will stay disciplined and focused.

Every department should be required to articulate how its plans link back to the organization's original digital transformation goals.

10.2 BUILD PERSONAL BRIDGES OF UNDERSTANDING AMONG DEPARTMENTS:

When two departments that have traditionally worked together in an informal way are suddenly asked to formalize and define their working relationships in ServiceNow, it can create friction and lead to disagreements. The relationships that were forged in a non-structured setting may suddenly break down in ServiceNow's highly regimented, clearly defined work environment. The solution is to focus on building personal bridges of understanding among departments. After all, behind every formal working relationship codified in the ServiceNow platform are human beings who both want to move forward in the best way they know how. Especially in the early stages of connecting departments in ServiceNow, you want to foster development of personal relationships by arranging face-to-face meetings, and encouraging departments to work as closely and collaboratively as possible to design their new, shared ServiceNow workflows.

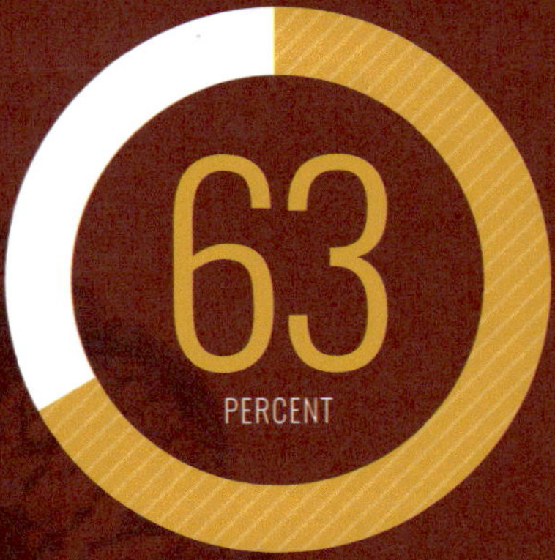

of managers believe the pace of digital transformation is too slow in their organization, primarily due to a perceived lack of urgency and poor communication about the benefits of upgrading.

MIT SLOAN MANAGEMENT REVIEW AND CAPGEMINI CONSULTING STUDY

10.3 JUST SAY NO TO POORLY CONCEIVED AND LEGACY WORKFLOWS:

ServiceNow only fires on all cylinders when the workflows you build into ServiceNow are optimized for this highly evolved system. Unfortunately, not every department will be able to design these optimized workflows. Some departments will attempt to force their legacy workflows to adapt to the ServiceNow model with minimal changes. Other departments will devise poorly conceived workflows that will lead to a world of hurt down the road. You must be prepared to identify these deficient implementation plans and send their creators back to the drawing board; your professional ServiceNow implementation consultant can play a key role in helping with this process. If you allow departments to make the final calls on what they think is best, you're setting up your organization for long-term bumps in the road.

It's important to push back on poorly conceived ServiceNow workflows that you'll live to regret.

10.4 ALLOCATE MORE RESOURCES THAN YOU THINK YOU'LL NEED:

As you prepare for your journey with ServiceNow, you must budget a certain level of resources over a certain timeframe to support the initiative. No matter how many resources you budget for, you should assume you'll need more. As the various departments and planning teams within your organization get going, they will no doubt get more and more excited about ServiceNow's potential, developing innovative ideas for further transforming workflows and business processes. As long as these ideas can be executed in ways that conform to ServiceNow best practices, you don't want to have to turn any of them down for budgetary reasons. After all, once the enthusiasm for change has dissipated, you may never get it back. You can proactively plan for this increased demand by creating and maintaining a long-term ServiceNow implementation roadmap. When you're looking three to five years down the road, you're far more likely to be accurate in anticipating ServiceNow demand.

10.5 LISTEN TO THE WISDOM OF YOUR IMPLEMENTATION CONSULTANT:

When organizations retain the services of a ServiceNow implementation consultant, they tend to view the consultant's role as the implementer, not the visionary and leader. These organizations erroneously believe that the consultant, coming in from the outside, cannot possibly understand the needs and priorities of the organization as well as its internal leadership does. This is a self-destructive line of thinking. Integration consultants are trained to shine a spotlight on client-developed implementation plans that are unnecessarily complex and unwieldy, and push back against poorly conceived workflows. Even when you fundamentally disagree with your implementation consultant's recommendations, you should aim to find a compromise that you both can get behind. Your implementation consultant's goal is to get you to the best possible long-term outcomes, and will draw upon years of wisdom and on-the-ground experience to get you there. Take your consultant's advice to heart.

You cannot afford to ignore or override the advice of your ServiceNow implementation consultant.

As with any new platform implementation, integrating and coordinating ServiceNow implementation across many disparate business units and functions in a protracted timeframe can feel overwhelming and stressful. But the end result is an organization that is digitally transformed from the inside out. To ensure this digital transformation proceeds as smoothly as possible, you want to require every department to link its plans back to enterprise-wide priorities, build personal bridges of understanding among departments, reject poorly conceived and legacy workflows as incompatible with ServiceNow, allocate more resources than you think you'll need, and give deference to the wisdom of your implementation consultant.

2 OUT OF 3

business leaders believe that their companies must pick up the pace of digitalization to remain competitive.

47%

of CEOs are being challenged by the board of directors to make progress in digital business.

56%

said that their digital improvements have already improved profits.

Although more CEOs have digital ambitions, the survey revealed that nearly half of CEOs have no digital transformation success metric.

Final Thoughts

Digital transformation is about much more than technical competency and carefully laid implementation plans. It is also about telling engaging and informative stories—the type of stories that get your organization excited about reimagining workflows and business processes. For all the time you spend communicating in KPIs and software specs and FTE allocations, you also want to communicate in anecdotes and metaphors and diagrams. You want to convince department leaders and frontline employees alike that ServiceNow's Enterprise Service Management (ESM) model will support their specific needs, priorities, and goals, even as ServiceNow simultaneously drives enterprise-wide efficiencies and bottom-line profitability.

Digital transformation succeeds when you get stakeholders excited about reimagining their future.

Digital transformation does not come easy for organizations that have traditionally let individual departments define and manage their own workflows. Disconnected, fragmented business processes are all that many departments have ever known and, as if through instinctive survival mechanism, these departments cling tightly to the status quo. To get the pendulum swinging in the opposite direction, departments must be reassured that it's in their best interest to give up some control and autonomy. They need to know that whatever they give up, they'll gain back a hundred times over, in the form of valuable automations and insightful analyses and interdepartmental collaborations.

Remember, your IT department cannot push through an enterprise transformation at the scale of ServiceNow on its own. Leaders from across all units and business processes must work together to create the game-changing digital transformation that your organization deserves. Each stakeholder and each team must do its part to become a strong, enduring link in the digital transformation chain.

Your digital transformation chain is only as strong as its weakest link.

Afterword

As you get the ball rolling on your digital services transformation using ServiceNow, you're going to quickly feel buried under the weight of scoping meetings, progress reports, custom coding requests, unanticipated bottlenecks, and other routine (and non-routine) implementation challenges. You may have no problem staying calm and focused under pressure, but you are responsible for maintaining more than just your own composure and momentum. It's your job to to keep the universal message of ServiceNow's value proposition alive and well across the many discrete departments and divisions of your organization. You need to continuously remind your stakeholders that ServiceNow is a top priority worthy of the entire organization's sustained focus and effort. Let's explore four essential tips for keeping everyone working in earnest toward your big-picture digital services transformation goals:

It's your responsibility to keep the promise of ServiceNow alive and well across your organization.

1. APPOINT EVANGELISTS TO GET OUT THE WORD:

On your own, your voice can only be amplified so much. But when you enlist hand-picked evangelists to educate others on the value of ServiceNow, you'll reach a much wider audience. Evangelists are influencers within your organizations whose opinions others already listen to and trust. That's why you want to routinely brief your evangelists on your progress, incremental wins, and future plans. When you get them excited, you can count on them to help spread this excitement to everyone in their spheres of influence.

2. VERIFY ALL NEW FEATURES WORK AS DESIGNED:

Like it or not, you're going to be confronted with skeptics and critics who won't buy into ServiceNow transformative power, no matter what you do. While you may not be able to change their minds during your ServiceNow implementation journey, you absolutely cannot afford to lend fuel to their fire. In other words, you've got to prioritize making sure that all of ServiceNow's apps and features work as intended by the time they're rolled out to the end user. If your rollouts are plagued by technical glitches and other setbacks, all of the effort that you've put into extolling ServiceNow's virtues will be wasted. The average end user will form an impression of ServiceNow based almost entirely on their initial first experience using it. It's up to you to make their first experience an overwhelmingly positive one.

3. GROW AND EVOLVE IN LOCKSTEP WITH YOUR END USERS:

As your end users incrementally get up to speed on the ServiceNow platform, you don't want ServiceNow to become a static, stale experience. Especially as you—and they—come up with new ideas for how to grow and evolve the platform over time, it is incumbent on you to capitalize on this organically generated momentum. You want to respond as quickly as possible with solutions and updates that will keep them engaged, feeling that their input is valued, and coming back for more. ServiceNow is a dynamic, fully responsive platform that can drive all manner of continuous process improvements; it's up to you to capitalize on each of these opportunities.

> **During your ServiceNow rollout, it's your responsibility to keep end users engaged, feeling valued, and coming back for more.**

4. SHOW—DON'T TELL—STAKEHOLDERS WHY CHANGE IS GOOD:

When you extol the virtues of ServiceNow in meetings with stakeholders, you may forget that your audience has "heard it all before." In other words, your sales pitch promising modernization and streamlining with this powerful platform is something your audience has been told repeatedly about completely unrelated projects—projects that have experienced varying degrees of success. To prove to your audience that you can deliver on your ambitious vision and promises, you can't just keep telling them what you plan to do—you've got to show them. Instead of running through PowerPoint slides during a meeting, you want to schedule a webinar where you demo a particular series of ServiceNow features. Instead of telling your audience about ServiceNow's strong ROI in a particular area, you want to show them actual data about what ServiceNow clients have achieved, and relate it back to your own ROI projections. Instead of telling them how much time they'll save with ServiceNow, you want to invite your end users to document how long they spend completing a task in ServiceNow vs. with their legacy workflows. By baselining legacy workflows against ServiceNow workflows, you're telling a compelling story that matters to ordinary end users.

As you become immersed in the technical and procedural details of ServiceNow, don't forget about your audience of stakeholders and end users who aren't immersed. To create ServiceNow value for them, and to spread ServiceNow victories to the many departments and divisions of your enterprise, you need to focus on maintaining positive momentum for everyone. This momentum-building journey starts with hand-picking evangelists to help get out the word, verifying that every app and feature works as intended, growing and evolving in lockstep with your end users' journey, and demonstrating to your stakeholders why they should believe in the potential and promise of ServiceNow.

5. SHARE YOUR IMPLEMENTATION ROADMAP FAR AND WIDE:

One of the biggest mistakes that businesses make during ServiceNow implementation is to hold their implementation roadmap too close to the vest. Although it takes time to educate stakeholders about the three- to five-year plan for rolling out ServiceNow, this outreach is essential. End users need to understand the big picture for where the enterprise is headed—not only so that they can prepare their own workflows and processes for the changes ahead, but also so that they can get excited and buy into the ServiceNow vision. When developed correctly, your ServiceNow implementation roadmap serves as the link between day-to-day workflows and the organization's strategic, long-term goals. To maximize its impact, you want to share your roadmap as often and as widely as possible.
As you become immersed in the technical and procedural details of ServiceNow, don't forget about your audience of stakeholders and end users who aren't immersed. To create ServiceNow value for them, and to spread ServiceNow victories to the many departments and divisions of your enterprise, you need to focus on maintaining positive momentum for everyone. This momentum-building journey starts with hand-picking

evangelists to help get out the word, verifying that every app and feature works as intended, growing and evolving in lockstep with your end users' journey, demonstrating to your stakeholders why they should believe in the potential and promise of ServiceNow, and sharing your ServiceNow implementation widely.

About Crossfuze

Crossfuze is a ServiceNow Gold Partner with the highest-ranking customer satisfaction score worldwide. With a global team that has completed over 1,000 implementations, Crossfuze is a proven leader in platform development and support, as well as the delivery of specialized solutions including IT Operations Management, IT Asset Management, IT Service Management, SecOps, and Customer Service Management. Crossfuze clients can also take advantage of consulting services to develop a ServiceNow vision, business case, and win-worthy implementation roadmap. With focused offerings and 12-plus ServiceNow platform-enhancing Quickfuze™ applications and integration solutions, Crossfuze is uniquely positioned to help clients deliver long-term ServiceNow success with speed, agility, and confidence.

The *10 Pillars of ServiceNow Digital Transformation for the Non-IT Leader* is the third volume in Crossfuze's 10 Pillars of ServiceNow Success book series. To download copies of the other volumes, visit Crossfuze's ServiceNow Resource Center, or request a printed copy at **LetsTalk@crossfuze.com.** In our ServiceNow Resource Center, you'll also find lots of other insights, information, and tips that will further ServiceNow success across your organization.

Crossfuze ServiceNow Resource Center:
https://esmresources.crossfuze.com

Made in the USA
Monee, IL
05 November 2026

67384982R00059